Practical Spreadsheet Risk Modeling for Management

Practical Spreadsheet Risk Modeling for Management

Dale Lehman
Huybert Groenendaal
Greg Nolder

CRC Press
Taylor & Francis Group
Boca Raton London New York

CRC Press is an imprint of the
Taylor & Francis Group an **informa** business
A CHAPMAN & HALL BOOK

Chapman & Hall/CRC
Taylor & Francis Group
6000 Broken Sound Parkway NW, Suite 300
Boca Raton, FL 33487-2742

International Standard Book Number: 978-1-4398-5552-2 (Hardback)

Library of Congress Cataloging-in-Publication Data

Lehman, Dale E.
Practical spreadsheet risk modeling for management / Dale Lehman, Huybert Groenendaal, and Greg Nolder.
p. cm.
ISBN 978-1-4398-5552-2 (alk. paper)
1. Risk management. 2. Risk management--Mathematical models. 3. Electronic spreadsheets. I. Groenendaal, Huybert. II. Nolder, Greg. III. Title.

HD61.L44 2012
658.15'50285554--dc22 2011012291

Visit the Taylor & Francis Web site at
http://www.taylorandfrancis.com

and the CRC Press Web site at
http://www.crcpress.com

For my parents, Pearl and Sid; my wife, Nancy; and my son, Jesse

—D. L.

For my parents, Huib and Greet; my wife, Rebecca;
and my daughters, Lilian and Violet

—H. G.

Dedicated to my wife, Rachel; daughters, Liza, Laura, and Jenna; parents, Gary Nolder, Ann Heinz, and Jerry Heinz; parents-in-law, Richard and June Keeley; sister, Jennifer Nolder; grandparents, Galen and Elizabeth Nolder; and grandparents, Dewitt and Eda Walters

To my extended family: Debbie, Chelsea, Geoff, Keeley, and Caroline Krombach; C. K. Krombach; Geoff Packard; Tim, Pam, Jason, Adam, and Alex Ford; Mike, Becky, Anna, and Erin Shukis; Joe, Julia, Tom, Maggie, and Will Heinz; Scott and Zach Heinz; Bill, Heather, and Tracy Nolder; Vic, Connie, Scot, Laura, and Danica Walters; Homer and Dorlene Nolder; Bernard and Helene Anderson; Matthew and Helen Bach; and Robert, Turid, Arne, and Tove Anderson

—G. N.

Contents

Preface ... xi
Acknowledgments ... xiii
Introduction ... xv
The Authors ... xix

1. Conceptual Maps and Models ... 1
- 1.1 Introductory Case: Mobile Phone Service ... 1
- 1.2 First Steps: Visualization ... 3
- 1.3 Retirement Planning Example ... 5
- 1.4 Good Practices with Spreadsheet Model Construction ... 10
- 1.5 Errors in Spreadsheet Modeling ... 13
- 1.6 Conclusion: Best Practices ... 15

2. Basic Monte Carlo Simulation in Spreadsheets ... 23
- 2.1 Introductory Case: Retirement Planning ... 23
- 2.2 Risk and Uncertainty ... 24
- 2.3 Scenario Manager ... 25
- 2.4 Monte Carlo Simulation ... 27
- 2.5 Monte Carlo Simulation Using ModelRisk ... 30
- 2.6 Monte Carlo Simulation for Retirement Planning ... 39
- 2.7 Discrete Event Simulation ... 47

3. Modeling with Objects ... 51
- 3.1 Introductory Case: An Insurance Problem ... 51
- 3.2 Frequency and Severity ... 52
- 3.3 Objects ... 59
- 3.4 Using Objects in the Insurance Model ... 60
- 3.5 Modeling Frequency/Severity without Using Objects ... 65
- 3.6 Modeling Deductibles ... 69
- 3.7 Using Objects without Simulation ... 76
- 3.8 Multiple Severity/Frequency Distributions ... 77
- 3.9 Uncertainty and Variability ... 82

4. Selecting Distributions ... 91
- 4.1 First Introductory Case: Valuation of a Public Company—Using Expert Opinion ... 91
- 4.2 Modeling Expert Opinion in the Valuation Model ... 94

4.3 Second Introductory Case: Value at Risk—Fitting Distributions to Data ... 99
4.4 Distribution Fitting for VaR, Parameter Uncertainty, and Model Uncertainty ... 99
4.5 Commonly Used Discrete Distributions ... 111
4.6 Commonly Used Continuous Distributions ... 113
4.7 A Decision Guide for Selecting Distributions ... 116
4.8 Bayesian Estimation ... 117

5. Modeling Relationships ... 129
5.1 First Example: Drug Development ... 129
5.2 Second Example: Collateralized Debt Obligations ... 134
5.3 Multiple Correlations ... 139
5.4 Third Example: How Correlated Are Home Prices?—Copulas ... 141
5.5 Empirical Copulas ... 146
5.6 Fourth Example: Advertising Effectiveness ... 148
5.7 Regression Modeling ... 150
5.8 Simulation within Regression Models ... 154
5.9 Multiple Regression Models ... 159
5.10 The Envelope Method ... 161
5.11 Summary ... 164

6. Time Series Models ... 171
6.1 Introductory Case: September 11 and Air Travel ... 171
6.2 The Need for Time Series Analysis: A Tale of Two Series ... 172
6.3 Analyzing the Air Traffic Data ... 175
6.4 Second Example: Stock Prices ... 183
6.5 Types of Time Series Models ... 188
6.6 Third Example: Oil Prices ... 190
6.7 Fourth Example: Home Prices and Multivariate Time Series ... 196
6.8 Markov Chains ... 200
6.9 Summary ... 205

7. Optimization and Decision Making ... 211
7.1 Introductory Case: Airline Seat Pricing ... 211
7.2 A Simulation Model of the Airline Pricing Problem ... 212
7.3 A Simulation Table to Explore Pricing Strategies ... 216
7.4 An Optimization Solution to the Airline Pricing Problem ... 219
7.5 Optimization with Simulation ... 226
7.6 Optimization with Multiple Decision Variables ... 229
7.7 Adding Requirements ... 231
7.8 Presenting Results for Decision Making ... 235
7.9 Stochastic Dominance ... 238
7.10 Summary ... 244

Appendix A: Monte Carlo Simulation Software 251
A.1 Introduction 251
A.2 A Brief Tour of Four Monte Carlo Packages 252

Index 255

Preface

This book is about building spreadsheet models to support decisions involving risk. There are a number of excellent books on this subject, but we felt the need for (1) a book that incorporates some of the latest techniques and methods in risk modeling that are not covered by other books, and (2) a book aimed at practitioners, with many practical real-world examples. Risk analytics is a rapidly developing field, and we have strived to present examples that illustrate the power of numerous techniques and methods that are not yet very common. Our experience with people who need to use these tools "in the field" is that they need material that is theoretically sound as well as practical and straightforward enough for them to utilize in their work. To this end, we have dispensed with the use of complex mathematics and concentrated on demonstrating how powerful techniques and methods can be used correctly within a spreadsheet-based environment to help make decisions under risk.

Whom Is This Book for?

This book is written for anyone interested in conducting applied risk analysis. This applies to analysis in business, engineering, environmental planning, public policy, medicine, or virtually any field amenable to spreadsheet modeling. If you intend to use spreadsheets for purposes of decision-supporting analysis, rather than merely as placeholders for numbers, then it is likely appropriate to include risk analysis in your spreadsheets. Consequently, this book may appeal to business students, students using quantitative analysis in other fields, or applied practitioners who use spreadsheet models.

This book is written at a beginner to intermediate level appropriate for graduate students or advanced undergraduates interested in risk analysis modeling. We have kept the mathematics to a minimum; more technical descriptions of many topics are available from a number of reference sources, but minimal mathematical background is required in order to use this book. We do assume that the reader is familiar with how to use Microsoft Excel® and has an understanding of basic statistical concepts (such as mean, standard deviations, percentiles, and confidence intervals). Readers without this background will probably, from time to time, need to supplement this text with additional materials.

This book is suitable for use in courses ranging from a few days to an entire semester. For a short course, some of the material can be preassigned, and the end-of-chapter problems can be used to practice the techniques covered within the chapter. For longer courses, the entire set of end-of-chapter problems, which spans a range of industries from health to finance (as explained in the next section), can be covered. Because this text does keep mathematical language to a minimum, instructors can supplement it with additional materials that stress the mathematical background and foundations. In that case, the current text can be used for its case studies or it can be embedded in broader courses about decision making. We have also found it to be valuable to add a project-based component after covering the material in the book.

This book uses the ModelRisk® software throughout and comes with a 120-day trial version of the software, which is available from the website http://www.vosesoftware.com/lehmanbook.php. However, the focus of the book is not on software, but rather on the different techniques and methods in risk analysis that can be used to build accurate and useful models to support decision making. We regularly use many different software packages within our work, each with its own advantages and disadvantages. Users of any spreadsheet-based Monte Carlo simulation software products (other than ModelRisk) will find that much of the material applies equally well to them. In Appendix A we include a brief overview of the four major Monte Carlo simulation Excel add-in software packages available. A number of techniques and methods in the text are directly available only in ModelRisk, but can typically also be implemented within the other software packages with some modification.

What Additional Features Supplement This Book?

A number of features are available to supplement the text. The book website (http://www.epixanalytics.com/lehman-book.html) has links to all of the following:

- The spreadsheet models used in the text
- A few short videos showing how to use many of the features of ModelRisk
- A Wiki site to permit uploading of further examples and solutions
- Text errata

Instructors adopting the text for course use will find teaching materials at http://www.crcpress.com/product/isbn/9781439855522. These include an instructor guide and solutions for the exercises in the book.

Acknowledgments

Considerable thanks are due to my students over the years. In particular, Jennifer Bernard, Gavin Dittman, Mark Giles, Paul Hitchcock, Leona Lien, Phong Moua, Ronald Paniego, Antonia Stakhovska, and Katherine Tompkins helped improve the clarity and coverage of the text. I owe a lot to my coauthors, Huybert Groenendaal and Greg Nolder, who have my respect as two of the best risk analysts around. I was perfectly capable of making my own errors, but would not have gotten past these without Huybert and Greg's collaboration. Thanks also to the people at Vose Software for producing the excellent ModelRisk software and making a trial version available for this book. Finally, this book could not have been undertaken (and completed) without the support of my wife and son, who put up with the unpredictable hours and temperament of an author.

—Dale Lehman

My thanks go out to many more people than I can mention here, so I'll keep it brief. Dale and Greg, thanks for being such great coauthors and for keeping the book very practical and hands-on. Thanks to all my colleagues and peers in the field of risk analysis and risk modeling over the years, from whom I've learned much. Last but not least, I am grateful to all our consulting, training, and research clients with whom I've had the privilege to work over the years. There are few things as professionally satisfying as collaborating with interesting and diverse people on complex and challenging risk analysis projects addressing real-life problems.

—Huybert Groenendaal

Special thanks to Dale Lehman and Huybert Groenendaal for including me in the development of this book, as well as to Huybert and our colleague Francisco Zagmutt, from whom I have learned much about risk analytics. Vose Software has been extremely supportive of our efforts, so many thanks to David Vose, Timour Koupeev, and Stijn Vanden Bossche.

Thanks to Dale Fosselman at Denali Alaskan Federal Credit Union for all the support as well as to my Denali colleagues Bob Teachworth, Eric Bingham, Joe Crosson, Lily Li, Keith Bennett, Pam Gregg, Jonathan Soverns, Mike Gordon, Bill Boulay, Robert Zamarron, and Saundra Greenwald.

Finally, thanks to the many people, whether named here or not, that I have the privilege to know, including Stuart, Erin, Maureen, Madison, and Emily Wright; Josh, Heather, Hailey, and Tyler Matthews; St. Luke's United Methodist Church; Janet Forbes; Fred Venable; Jim and Leigh Ramsey; Kay Coryell; Lynda Fickling; Dave, Elizabeth, Brad, and Erin Laurvick; Steve and Martha Riley; Chris Wilterdink; Susan Johnson; Bob and Sharon Oliver; Paul Peterman; Scott Kohrs; Brenda Schafer; Dave Cupp; Barry Curtis; Karin Wesson; Jenna Wilcox; Sallie Suby-Long; Cindy Raap; Steve, Cathy, and Elise Collins; Ken Fong; Tim Boles; Shawn Slade; Dick Frame; Ken Whitelam; Julia Murrow; Dick Evans; Kevin and Ruth Johaningsmeir; Ren, Trudy, Carah, and Kendall Frederics; Jay and Renee Carlson, Ron and Ruth Hursh; Dan Knopf; Cathy Strait; Jennifer Dierker; Art Hiatt; Bill Knowles; Red Rocks Circle; Jim and Shirley Lynch; Kevin Freymeyer; Barry Hamilton; Dave Mirk; Ann Barnes; Jeff Raval; Jacquelyn Pariset; Julianne Garrison; Tracey Ayers; Elizabeth Brownell; James, Lori, Monica, and Abbey Sigler; Stace Lind; Dana, Danna, Katy, Claire, Evan, Nathan, and Ben Nottingham; Yoram, Anat, Yarden, Almog, Agam, and Arbel Sharon; Ariel and Alexa; John, Susan, Emily, and Anna Barr; Ramon, Sandra, Dana, and Kyla Colomina; Mike, Carol, and Sydney Vestal; Barbara O'Neill; Tony Gurule; Michael Ruston; Chris, Stephanie, and Lyndon Burnett; Doug and Nancy Heinz; Phyllis and Ron Lemke; Jerry Ryan; David and Sharron Prusse; Otto and Ana Brettschneider; Ken and Joanne Raschke; Gwenne, Steve, and Sebastian; Karen and Don; Jim and Janet; Susie and Charles; Rand Winton; Ice Breeden; David Wandersen; Kimberly Sweet; Gheorghe Spiride; Tom Hambleton; Marc Drucker; David Hurwitz; First United Methodist Church; Drew Frogley; Gary Sims; Michael Johnson; Lynn Anderson; Cindy Gomerdinger; Zunis; Neal Westenberger; Forney, Mary Lou, and James Knox; Jason Reid; Jackie Herd-Barber; Jeff Brunner; John Johnston; Terry and Beth Schaul; Bill Fields; Ledo's; Napoleon, Kip, Rico, Pedro, Deb, Grandma, Rex, and LaFawnduh; Ram's Horn; Deb and Alex; Piney Valley Ranch; Trail Ridge Road; Val and Earl; Burt and Heather; Charlie, Cassie, Thunder, Rain, Shine, Digger, Copper, Cupcake, Sammy, Bambi, Blitzen, Earl, and Sooner.

—Greg Nolder

Introduction

Risk: What This Book Is About

Financial meltdown, oil spills, climate change: we live in a risky world. Alan Greenspan (2007) has called it "the age of turbulence."[*] A search of the Business Source Premier database for the subject terms "risk" and "uncertainty" yields the following number of citations over each of the past five decades:

- 1960s: 520 citations
- 1970s: 1,979 citations
- 1980s: 4,824 citations
- 1990s: 11,552 citations
- 2000s: 50,489 citations

Further evidence of the increasing attention paid to risk is shown in Figure 0.1, which tracks the use of the word "risk" in published books over the past 200+ years.[†]

References to risk were relatively constant from 1800 through 1960 but have increased markedly since then. This does not necessarily mean the world has become a riskier place. Indeed, in a very real sense, risk has always been part of the human experience. Arguably, risks for early humans were greater: Predators, natural catastrophes, and disease were more threatening and severe than they are today. So, we probably do not live in a uniquely risky age. But we do live in the age of *risk analysis*. Never before have so many people had access to the software tools to conduct sophisticated (as well as simple) risk analyses.[‡]

[*] Greenspan, A. 2007. *The Age of Turbulence*. New York: Penguin Press.

[†] These data come from the Google labs books Ngram Viewer, http://ngrams.googlelabs.com. Note that this source is not a random sample and it is not the complete population of published books. However, it constitutes a large database of published books and the pattern closely matches that found from other sources.

[‡] For an excellent history of risk analysis, see Bernstein, P. L. 1998. *Against the Gods: The Remarkable Story of Risk*. New York: John Wiley & Sons.

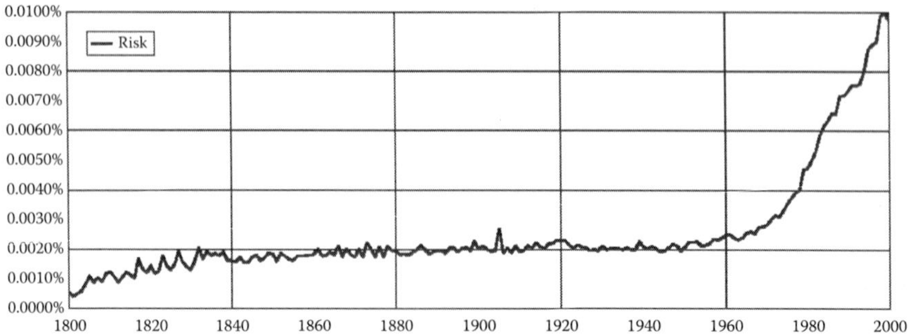

FIGURE 0.1
"Risk" references over the ages.

Primitive humans did not conduct sophisticated risk analyses. They mostly relied on instinct.* Modern risks—the products of complex technologies and economies—challenge us, and our instincts are often maladapted to making good decisions. In particular, human frailties when dealing with risky situations and probabilities have been well documented.† When someone is faced with complex modern risks, decision making cannot be left to instinct alone: A more systematic approach can provide great insight and better understanding to improve decision making.

At the same time, spreadsheets have become ubiquitous and are used in business, engineering, policy analysis, and virtually in any situation in which people analyze to support decisions. This book is about the marriage of these two evolutions: risk analysis and spreadsheet modeling. Our belief is that a number of tools are readily available that permit sophisticated risk analysis to be conducted in a variety of situations without a need for an extensive mathematical background.

How This Book Is Organized and How to Use It

The first three chapters comprise material covering how to construct spreadsheet models, how to integrate simulation modeling into spreadsheets, and

* Malcolm Gladwell (*Blink: The Power of Thinking without Thinking*, Back Bay Books, 2007) describes our abilities to react quickly under uncertainty. However, even when our instantaneous reactions serve us well, there are usually years of preparation, education, and training that permit our instincts to perform well at these times.

† The classic reference is Kahneman, D., P. Slovic, and A. Tversky. 1982. *Judgment under Uncertainty: Heuristics and Biases*. Cambridge, England: Cambridge University Press. A more comprehensive and recent reference source is Koehler, D. J. and N. Harvey, eds. 2004. *Blackwell Handbook of Judgment and Decision Making*. Hoboken, NJ: Wiley-Blackwell.

how to use ModelRisk Objects within spreadsheets to extend the capabilities of simulation models. Advanced readers can self-study the first two chapters, but all readers should read these three chapters. The remaining chapters (4–7) need not be covered in sequence or in entirety. Chapter 4 expands on the types of distributions that can be used in simulation modeling and also discusses fitting distributions to data. Chapter 5 focuses on estimating relationships between uncertain variables and on using simulation to represent the uncertainty about these relationships. Chapter 6 deals exclusively with time series data and forecasting. Chapter 7 examines optimization under uncertainty.

Also, there is an appendix that describes several spreadsheet simulation software packages (ModelRisk, Crystal Ball®, @Risk®, and RiskSolver®).

Each chapter has eight end-of-chapter problems. They span a range of industries, including

- Health care
- Transportation
- Finance and insurance
- Consumer/retail
- Technology
- Natural resources
- Manufacturing
- Sports and entertainment

This forms a matrix of seven chapters by eight sectors. Readers interested in a particular area can focus on the chapter problem devoted to that sector. We believe that risk analysis requires repeated application to novel situations, so these problems are essential to learning how to conduct practical risk analysis. We also believe that the diversity of these examples will help develop modeling skills for all of these sectors, so we encourage readers to utilize as many of these problems as time permits.

Before attempting the end-of-chapter problems, readers should verify that they comprehend the material in each chapter. The best way to do this is to reproduce the analyses shown in the text, using the spreadsheet models from the book's website.*

* In theory, all simulation results in the book can be reproduced precisely by running 10,000 simulations with a manual random seed value equal to zero. It is possible that use of a different version of Excel or ModelRisk, or slight differences in the layout of a spreadsheet, may cause the results to diverge, but the differences should be minor.

The Authors

Dale Lehman is professor of economics and director of the MBA program at Alaska Pacific University. He also teaches courses at Danube University and the Vienna University of Technology. He has held positions at a dozen universities and at several telecommunications companies. He holds a BA in economics from SUNY at Stony Brook and MA and PhD degrees from the University of Rochester. He has authored numerous articles and two books on topics related to microeconomic theory, decision making under uncertainty, and public policy, particularly involving telecommunications and natural resources.

Huybert Groenendaal is a managing partner and senior risk analysis consultant at EpiX Analytics. As a consultant, he helps clients using risk analysis modeling techniques in a broad range of industries. He has extensive experience in risk modeling in business development, financial valuation, and R&D portfolio evaluation within the pharmaceutical and medical device industries; he also works regularly in a variety of other fields, including investment management, health and epidemiology, and inventory management. He teaches a number of risk analysis training classes, gives guest lectures at a number of universities, and is adjunct professor at Colorado State University. He holds MSc and PhD degrees from Wageningen University and an MBA in finance from the Wharton School of Business.

Greg Nolder is vice president of applied analytics at Denali Alaskan Federal Credit Union. The mission of the Applied Analytics Department is to promote and improve the application of analytical techniques for measuring and managing risks at Denali Alaskan as well as the greater credit union industry. Along with Huybert, Greg is an instructor of risk analysis courses for Statistics.com. Prior to Denali Alaskan, he had a varied career, including work with EpiX Analytics as a risk analysis consultant for clients from numerous industries, sales engineer, application engineer, test engineer, and air traffic controller. Greg received an MS in operations research from Southern Methodist University as well as a BS in electrical engineering and a BS in aviation technology, both from Purdue University.

1

Conceptual Maps and Models

LEARNING OBJECTIVES

- Use visualizations of a situation to sketch out the logic for a model.
- Translate the visualization into a spreadsheet model.
- Develop good procedures for building and documenting spreadsheet models.
- Appreciate the prevalence of errors in spreadsheets and learn how to prevent them.

1.1 Introductory Case: Mobile Phone Service

Cellular telephone service (often called "mobile phone service" or "Handys") has been a spectacular success. After its invention at Bell Labs, AT&T commissioned McKinsey & Company in 1980 to forecast cell phone penetration in the United States. Their forecast for the year 2000: 900,000 subscribers. The actual figure was 109 million (and 286 million at the end of 2009 and still counting—with more than 4 billion worldwide). Part of the successful business model has been that the handsets are given to consumers for "free" in exchange for their signing of a long-term contract (usually 2 years). One of the major business decisions that cellular phone service providers must make is how much to subsidize the consumer's acquisition of the phone and what long-term contract terms to incorporate to recover these costs over time.

This business decision is a typical decision problem in several aspects. Myriad factors can influence what the optimal decision is, and a model can be useful to provide insight into the situation, even though, by definition, a model is a simplification of reality. In order to provide usable insight for the decision problem, the analyst must decide which factors are important to include in a model and which can be ignored. In this case, the decision concerns the degree to which the initial cell phone acquisition should be subsidized, and we will ignore factors that are not directly germane to this decision.

An initial model for this problem can be seen in Figure 1.1, which shows a simple model of the cell phone pricing decision that omits many factors

	A	B	C	D
1	Parameters		Notes	
2	Average phone cost	$100	Given parameter value	
3	Phone price	$0	Given parameter value	
4	Total market size	10,000	Given parameter value	
5	Price sensitivity	80	Given parameter value	
6	Average net revenue (per month)	$20	Given decision value	
7	Average contract length (months)	24	Given decision value	
8				
9	Calculations		Formula	
10	Profit per customer	$380	"= B5*B4 + (B3 – B2)"	
11	Number of customer	10,000	"= B4 – B5 + B3"	
12				
13	Results			
14	Total profit	$3,800,000	"= B9*B8"	

FIGURE 1.1
Basic cell phone pricing spreadsheet model.

but includes the essential relationship between the cell phone price and the number of subscribers (and, hence, total profits). We have assumed certain costs (to the service provider) of the phone, average monthly net revenues per subscriber, total market size, and a parameter that reflects consumers' sensitivity to the price they are charged for the phone. (Our assumption is that the higher the phone price is—i.e., the lower that the subsidy provided is—the smaller the number of subscribers will be. Each $1 increase in the price of the phone leads to an assumed reduction of 80 subscribers.) Figure 1.1 illustrates a particular set of decision values our service provider might consider.

The value of a spreadsheet model lies primarily in its capability to perform "what if" analyses. (Monte Carlo simulation, introduced in the next chapter, can be thought of as "what if" analysis "on steroids.") For example, if we change the price of the phone to $100 (no subsidy for subscribers), the total profit falls from $3,800,000 to $960,000 (mainly due to the reduction in subscribers from 10,000 to 2,000).*

But, how did we begin constructing this model? The answer is that we start with a visualization of the problem.

* Many people are more familiar with using spreadsheet calculations than with using spreadsheet *models*. The difference can be illustrated by an example: If we replace cell B11 in Figure 1.1 with the number 10,000 (rather than the formula), then it correctly computes the profits of $3,800,000 at the phone price of $0. However, the "what if" analysis of changing the phone price to $100 will show profits to increase to $4,800,000. An alternative model assumes that there is no consumer response to the change in the phone price. Such an alternative spreadsheet can correctly be used as a calculator, but does not produce sensible "what if" analyses. In other words, to do insightful "what if" analysis, a spreadsheet model must be constructed that captures the essential logic of the problem being studied.

1.2 First Steps: Visualization

Since the integrity and therefore usefulness of a spreadsheet model depend on having a clear conceptual map of the appropriate relationships in the model, the question is: Where do you begin? Every spreadsheet model at some point begins with a blank spreadsheet, and the novice modeler often begins with a blank stare. The answer, however, is not to begin in the spreadsheet; rather, it is to start on a piece of paper. It is best to create a visualization of the model you want to build. Various authors have given these visualizations different names: influence diagrams (Cleman and Reilly 2010[*]), influence charts (Powell and Baker 2009[†]), mental models or visual models (Ragsdale 2007[‡]), or causal loop diagrams (Morecroft 2007[§]). The intent of these tools is identical: to visualize the parts of the model and their relationships. We will refer to any of these as visualization tools, and they are a critical first step in model building. Indeed, if you are careful in constructing your visualization, it can serve as a blueprint of your model and can almost automate the building of your spreadsheet model.

There is no standardized form to these visualization tools. We will use a variant of the influence chart included in Powell and Baker (2009). The diagram distinguishes between the following elements, each with its own shape:

- Objectives: use hexagons.
- Decisions: use rectangles.
- Input parameters: use inverted triangles. If these are key uncertainties, use dashed lines for the border of the triangle.
- Calculations: use ovals.

Every shape should have one or more connectors to other shapes where these are appropriate. For example, our spreadsheet model for the cell phone pricing problem was based on the influence chart shown in Figure 1.2.

The chart is started on the right side with the objective(s). Our simple first model has one objective—total profit—but your model may contain more objectives if your problem has multiple items of interest. For example, if you are modeling staffing of a call center, you might be interested in analyzing the average time to process calls as well as the total cost of staffing and operating the call center.

Building the chart is typically best done by decomposing backward from the objective, one step at a time, until you only have parameters and decisions. Intermediate stages involve calculations. To calculate our cellular

[*] Clemen, R.T. and Reilly, T. 2004. *Making Hard Decisions with Decision Tools Suite Update Edition.*
[†] South-Western College Pub., Powell, S.G. and Baker, K.R. 2009. *Management Science: The Art of Modeling with Spreadsheets,* John Wiley & Sons.
[‡] Ragsdale, C.T. 2007. *Spreadsheet Modeling & Decision Analysis,* South-Western College Pub.
[§] Morecroft, J. 2007. *Strategic Modelling and Business Dynamics,* John Wiley & Sons.

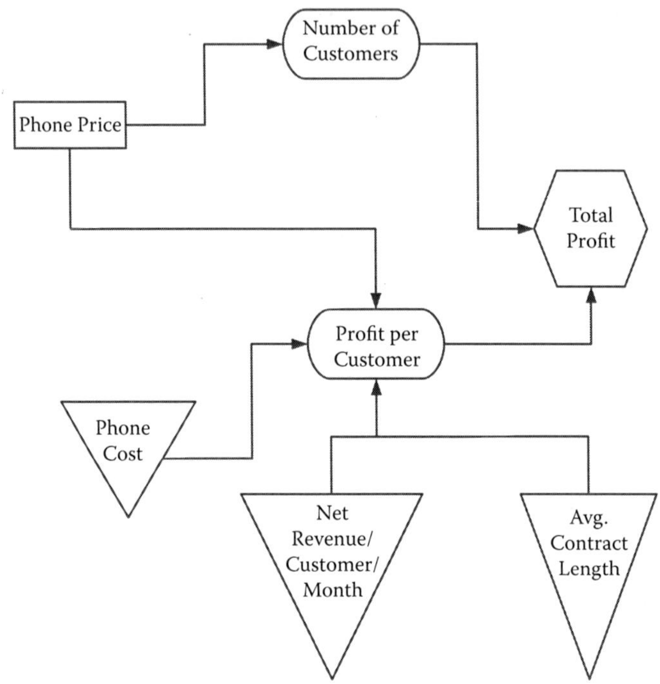

FIGURE 1.2
Basic cell phone pricing influence chart.

provider's profits, we need to know how many customers it has and what the average profit per customer is. Each of these is calculated based on other things (parameters) in the model. The number of customers is influenced by the handset price, the market size, and price sensitivity (respectively, one decision and two parameters). Average profit per customer is influenced by the handset price, the handset cost, average net profit per customer per month, and the average contract length (one decision and three parameters).

There are some important things to remember when constructing the visualization:

- Designation of something as a parameter does not mean it is obvious or trivial. It only means that your model will not attempt to explain where this variable comes from. Later models may well replace a parameter with some logic that shows the influences that determine such a parameter. But it is important to start simple with only the most important factors included in your model.
- Connector lines are very important. In most models, every parameter will directly or indirectly be linked to all objectives, but the connectors represent direct relationships between two elements in the visualization. If there is a connector, then the formula for that cell

in the spreadsheet should include reference to the cell to which it is linked. If there is no such connector, then there should not be a cell reference.

- Your visualization should not contain any numbers. You are trying to capture essential relationships, to help develop the structure of your model. Numbers belong in the spreadsheet, not in your visualization. A corollary to this is that, typically, you should not yet worry about lack of knowledge (even about key parameters) when building your visualization. It is always possible to develop some estimate for a parameter, and when none is available, Monte Carlo simulation (introduced and covered in the next chapter) is often a good way to model such uncertain values.*
- Try to be precise when defining terms in your visualization and ensure that they are measurable. For example, the objective "net present value of profits" is better than "profits" because it will more easily translate directly into your spreadsheet. Similarly, a decision of "handset subsidy" is easier to implement in the spreadsheet than a decision, "should we subsidize handsets?"

A more sophisticated version of our cellular model appears in Figure 1.3, where we have expanded on the fact that there is a time value of money and that demand sensitivity and average contract length are uncertain parameters. As a result, our spreadsheet model will contain some additional parameters and linkages (as discussed in more detail in the next chapter).

1.3 Retirement Planning Example

Additional care is required for appropriate visualizations for problems that involve a timed process. It is best to model the sequence of events or activities and their relationship explicitly over time. For example, suppose that we want to build a model of a typical retirement planning decision. Consider a 30-year-old with no current retirement fund beginning employment at an annual salary of $50,000. Suppose that this employee decides to set aside 10% of his or her pretax salary for a retirement fund, an amount to be matched by the employer (up to a maximum of 7%). Assume that he or she anticipates getting annual salary increases that exceed the inflation rate by 2%. (We will conduct the analysis in inflation-adjusted dollars, so as not to worry about

* This is not always true. Sometimes it is preferable to design a model that does not depend on unknown parameters—an alternative model (possibly a simpler model or a model that looks at the problem from a different perspective) where parameters with unknown values can be avoided.

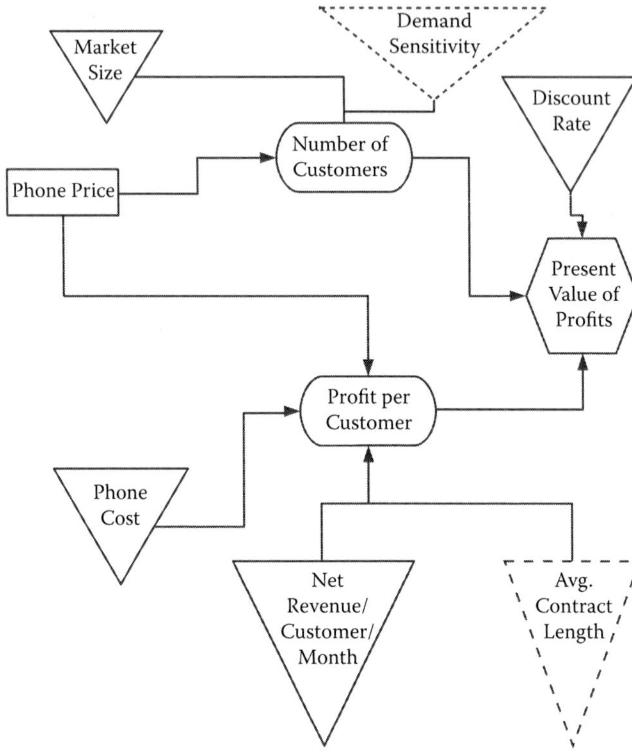

FIGURE 1.3
Enhanced cell phone pricing influence chart.

the value of a dollar in the future.) Our employee anticipates retiring at some age between 60 and 65 and wants to see how long the retirement funds will last under these assumptions and for each possible retirement age.

Let us further assume that our employee expects to get a return on retirement funds that averages 3% above the expected inflation rate and that after retirement he or she anticipates needing $50,000 per year (again, in inflation adjusted or real dollars) to live comfortably. Do not worry about whether these assumptions are correct at this point or even whether they are known for sure. It is the essential structure of the problem we need to capture. Figure 1.4 illustrates a visualization that captures the logic of our problem.

Note that no numbers appear in Figure 1.4. We have not indicated any objective and have not placed the initial fund balance in our diagram. We could do so, beginning with our worker at age 30 and moving through the entire lifetime; however, this would only clutter the diagram. What is important to capture in the visualization is the time structure of the problem and Figure 1.4 shows the recurring nature of the money going into and coming out of the retirement fund, which depends on our parameter assumptions and whether or not retirement age has been reached. Figure 1.4 provides

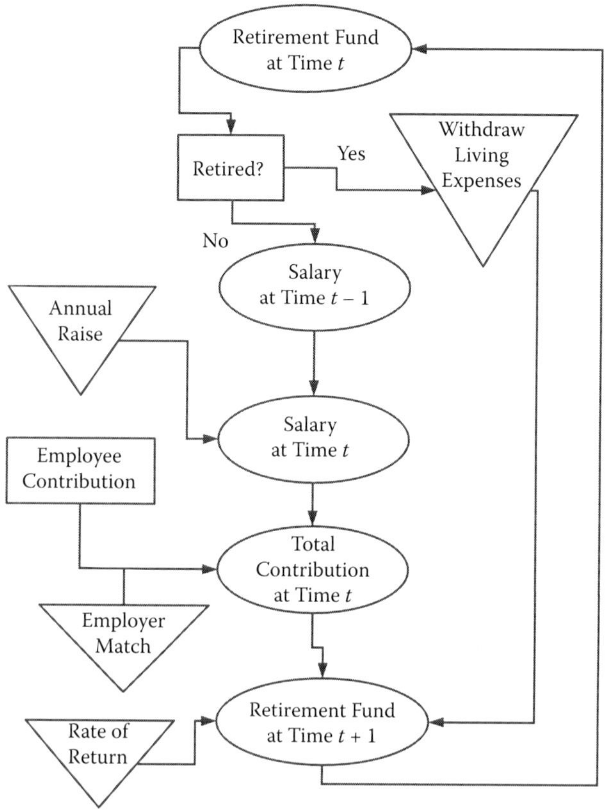

FIGURE 1.4
Retirement fund visualization.

a base for constructing a spreadsheet model shown in Figure 1.5. (The spreadsheet spans 70 years, but we have hidden most of the rows in the display; the complete spreadsheet is available on the book website.)

These are the salient features represented in this model, listed by the cell address:

A15: Rather than inserting the value 30 for our subject's current age, we have linked it to a parameter representing the current age. This permits the spreadsheet to be easily adapted to a different individual's circumstances. Similarly, the beginning fund balance, cell C15, can be easily changed in the parameter section without needing to change anything in the calculation section of the spreadsheet.

B15: This formula produces a zero or one value indicating whether or not the person has retired. It is calculated by comparing the current age with the chosen retirement age (which is a decision).

	A	B	C	D	E	F	G	H	I	J	K
1	**Parameters**										
2	Current Age	30									
3	Salary	50000									
4	Annual Raise	0.02									
5	Employee Contribution	0.1									
6	Employer Match	0.07									
7	Rate of Return	0.03									
8	Retirement Living Expenses	50000									
9	Initial Fund Balance	0									
10	**Decisions**		**Objective** (year funds run out)								
11	Retirement Age	65	=B2+SUM(K15:K85)								
12											
13	**Calculations**										
14	Age	Retired?	Fund at Start of Year	Return on Fund	Salary	Employee Contribution	Employer Match	Total Contributions	Withdrawals	Fund at End of Year	Fund still available?
15	=B2	=IF(A15>=$B$11,1,0)	=$B$9	=$B$7*C15	=(1-B15)*($B$3*(1+$B$4)^(A15-30))	=$B$5*E15	=IF($B$5>7%,7%,$B$5)*E15	=F15+G15	=B15*$B$8	=C15+D15+H15-I15	=IF(J15>0,1,0)
16	=A15+1	=IF(A16>=$B$11,1,0)	=J15	=$B$7*C16	=(1-B16)*($B$3*(1+$B$4)^(A16-30))	=$B$5*E16	=IF($B$5>7%,7%,$B$5)*E16	=F16+G16	=B16*$B$8	=C16+D16+H16-I16	=IF(J16>0,1,0)
17	=A16+1	=IF(A17>=$B$11,1,0)	=J16	=$B$7*C17	=(1-B17)*($B$3*(1+$B$4)^(A17-30))	=$B$5*E17	=IF($B$5>7%,7%,$B$5)*E17	=F17+G17	=B17*$B$8	=C17+D17+H17-I17	=IF(J17>0,1,0)
83	=A82+1	=IF(A83>=$B$11,1,0)	=J82	=$B$7*C83	=(1-B83)*($B$3*(1+$B$4)^(A83-30))	=$B$5*E83	=IF($B$5>7%,7%,$B$5)*E83	=F83+G83	=B83*$B$8	=C83+D83+H83-I83	=IF(J83>0,1,0)
84	=A83+1	=IF(A84>=$B$11,1,0)	=J83	=$B$7*C84	=(1-B84)*($B$3*(1+$B$4)^(A84-30))	=$B$5*E84	=IF($B$5>7%,7%,$B$5)*E84	=F84+G84	=B84*$B$8	=C84+D84+H84-I84	=IF(J84>0,1,0)
85	=A84+1	=IF(A85>=$B$11,1,0)	=J84	=$B$7*C85	=(1-B85)*($B$3*(1+$B$4)^(A85-30))	=$B$5*E85	=IF($B$5>7%,7%,$B$5)*E85	=F85+G85	=B85*$B$8	=C85+D85+H85-I85	=IF(J85>0,1,0)

FIGURE 1.5
Retirement spreadsheet model.

C15: Initially set the parameter for the starting fund. Thereafter, it equals the ending balance from column J on the prior row (e.g., C16 = J15).

D15: This calculates the return on the fund balance from the previous year, using the assumed rate of return. We are assuming that returns are earned at the end of each year.

E15: The annual salary is the product of the initial salary and the appropriate number of years of growth at the assumed annual raise. Note also that we multiply this by a factor (1-B15) that results in a salary of zero when the person is retired or the calculated salary when he or she is not retired. (The same result could be obtained with an appropriate IF function.)

F15, G15: These cells calculate the employee and employer contributions to the retirement funds, respectively. The employer's match is limited to equal the employee contribution, or 7%, whichever is lower. H15 simply sums these two contributions. Note that contributions will be zero after retirement since the salary will be zero.

I15: This cell calculates the postretirement withdrawals from the account. The use of the factor B15 in the formula ensures that no money is withdrawn prior to retirement since B15 = 0 at that time.

J15: The end-of-year fund is the sum of the initial fund, the annual return, and the employee and employer contributions during the year, minus the withdrawals (becomes the starting balance in the next year, C16).

K15: This cell indicates whether or not the end-of-year funds are positive. It is used to calculate the year that the retirement fund is exhausted in cell C11. The formula in C11 adds the initial age to the number of years that show a positive fund balance, thus yielding the age at which the funds are gone (our objective).

Note that all numbers appear in the parameter section and only formulas appear in the calculation section. This is an important good practice in spreadsheet modeling and permits the spreadsheet model to be adapted easily to varying circumstances without needing to reconstruct the model. Note also our use of relative and absolute cell addressing. Use of the dollar symbol ($) in front of a row number or column letter will freeze that number or letter; thus, it will not change when copied (called *absolute* addressing). Absence of the $ means that the row number will be changed when copied downward or the column letter will be changed when copied rightward in the spreadsheet (called *relative* addressing). Appropriate use of absolute and relative addressing greatly simplifies the construction of models by permitting many formulas to be entered once and copied throughout the spreadsheet. The flip side is that improper application of relative and absolute addressing is also one of the most common errors made in building spreadsheet models.

The value of this model lies in its ability to be used for a variety of "what if" analyses. The base case for our model is shown in Figure 1.6, which produces an age of 83 for when the retirement fund is exhausted if the person retires at age 65.

Our model can be used to examine the sensitivity of the exhaust year to our two decisions: the employee annual contribution to the fund and the chosen retirement age. We do this by using the Data Table command in Excel (found on the Data ribbon under "What If Analysis," as shown in Figure 1.7). The result is the table shown in Figure 1.8.

As Figure 1.8 shows, the retirement funds are expected to run out as early as age 66 (for the earliest retirement age and the lowest employee contribution factor) or as late as age 91 (if the worker works until age 65 and contributes 15% of annual salary). If our worker wishes to have the fund last until the expected lifetime (say 83 years old, for a female in the United States), then this result can be obtained by working until age 64 with a contribution rate of 12% (or working until age 63 with a contribution rate of 14%).

Our model is quite flexible and can produce a number of insights, but it is severely limited by the fact that there are a number of uncertainties not represented in the model. This is a good application of Monte Carlo simulation, and thus we will begin the next chapter with further refinements to this retirement model.

1.4 Good Practices with Spreadsheet Model Construction

If you are careful in constructing the visualization, building the spreadsheet model will be much easier and efficient. Organization is provided by the shapes of the boxes; put parameters in one section of the spreadsheet (possibly divided into those that are certain and those that are uncertain), decisions in a second section, calculations in the third section, and objectives in the final output section (frequently put near the top for ease of viewing). If the spreadsheet is large, then these may be placed on separate worksheets, and different portions of the calculations may require separate worksheets. However, when building a spreadsheet model, you should always try to start small. It is easy to expand or enhance a model once a simple version is working.

Typically, spreadsheet models would contain only numbers and labels in the parameter and decision sections (or inputs for probability distributions for uncertain parameters, as discussed in the next chapter). The calculations and objectives would consist purely of formulas—with no numbers whatsoever, only cell addresses. Any time that you want to use a different set of input parameters, you simply replace its value in the parameter section and it will then be used (because of the linking) in the calculation part of the

	A	B	C	D	E	F	G	H	I	J	K
1	**Parameters**										
2	Current Age	30									
3	Salary	$50,000									
4	Annual Raise	2%									
5	Employee Contribution	10%									
6	Employer Match	7%									
7	Rate of Return	3%									
8	Retirement Living Expenses	$50,000									
9	Initial Fund Balance	$0									
10	**Decisions**		**Objective** (year funds run out)								
11	Retirement Age	65	83								
12											
13	**Calculations**										
14	Age	Retired?	Fund at Start of Year	Return on Fund	Salary	Employee Contribution	Employer Match	Total Contributions	Withdrawals	Fund at End of Year	Fund still available?
15	30	0	0	0	50,000	5,000	3,500	8,500	0	8,500	1
16	31	0	8,500	255	51,000	5,100	3,570	8,670	0	17,425	1
65	80	1	147,976	4,439	0	0	0	0	50,000	102,415	1
66	81	1	102,415	3,072	0	0	0	0	50,000	55,488	1
67	82	1	55,488	1,665	0	0	0	0	50,000	7,152	1
68	83	1	7,152	215	0	0	0	0	50,000	-42,633	0
69	84	1	-42,633	-1,279	0	0	0	0	50,000	-93,912	0
70	85	1	-93,912	-2,817	0	0	0	0	50,000	-146,729	0
71	86	1	-146,729	-4,402	0	0	0	0	50,000	-201,131	0
72	87	1	-201,131	-6,034	0	0	0	0	50,000	-257,165	0
73	88	1	-257,165	-7,715	0	0	0	0	50,000	-314,880	0
74	89	1	-314,880	-9,446	0	0	0	0	50,000	-374,326	0
75	90	1	-374,326	-11,230	0	0	0	0	50,000	-435,556	0
76	91	1	-435,556	-13,067	0	0	0	0	50,000	-498,623	0
77	92	1	-498,623	-14,959	0	0	0	0	50,000	-563,582	0
78	93	1	-563,582	-16,907	0	0	0	0	50,000	-630,489	0
79	94	1	-630,489	-18,915	0	0	0	0	50,000	-699,404	0
80	95	1	-699,404	-20,982	0	0	0	0	50,000	-770,386	0
81	96	1	-770,386	-23,112	0	0	0	0	50,000	-843,497	0
82	97	1	-843,497	-25,305	0	0	0	0	50,000	-918,802	0
83	98	1	-918,802	-27,564	0	0	0	0	50,000	-996,366	0
84	99	1	-996,366	-29,891	0	0	0	0	50,000	-1,076,257	0
85	100	1	-1,076,257	-32,288	0	0	0	0	50,000	-1,158,545	0

FIGURE 1.6
Retirement spreadsheet calculation results.

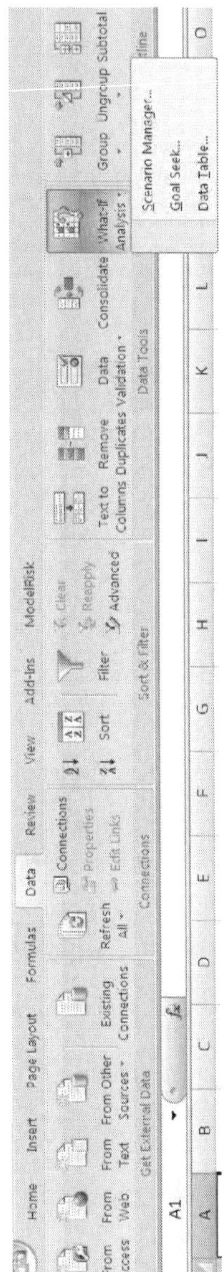

FIGURE 1.7
What If Analysis on the toolbar.

	A	B	C	D	E	F	G	H	I	J	K	L	M
1							EmployeeContribution Factor						
2		83	5.00%	6.00%	7.00%	8.00%	9.00%	10.00%	11.00%	12.00%	13.00%	14.00%	15.00%
3	Retirement Age	60	66	68	70	70	71	72	73	74	75	76	77
4		61	68	70	71	72	73	74	75	76	77	78	80
5		62	69	71	73	74	75	76	77	78	80	81	82
6		63	71	73	75	76	77	78	79	81	82	83	85
7		64	72	75	77	78	79	80	82	83	85	86	88
8		65	74	76	79	80	81	83	84	86	87	89	91

FIGURE 1.8
Data Table for the retirement model.

model. The advantage of this is that you can easily perform "what if" analyses by modifying the parameter values without worrying about whether you have appropriately modified the formulas. It also allows you (and other users of the model) to review your models much more quickly and easily for accuracy, logic, and input assumptions.

Beginning with a careful visualization and following these rules will greatly help the task of building a sound spreadsheet model. We do not mean to suggest it is always easy because there is often still a significant task in finding the right formulas to use in the calculations and identifying the correct structure for your visualization. The latter is what makes spreadsheet modeling as much an art as a science. As with any map, the usefulness of a model depends on the uses to which it will be put and the available information. One of the best ways to learn how to do this is through repeated application and trial and error. Beginning with a visualization often cures the problem of the blank stare or the question, "Where do I begin?". In addition, there are some ground rules to help you prevent a lot of common errors.

1.5 Errors in Spreadsheet Modeling

Spreadsheets have become a ubiquitous business tool. The first spreadsheet program, VisiCalc, was released in 1979. Lotus 123, released in 1983, became the first dominant spreadsheet program, but Excel quickly became the standard after its release. By 1989, it was the dominant spreadsheet program. As its use has grown, so has realization that many (indeed, most) spreadsheets in use contain errors, many of which are serious enough to have resulted in decisions that differ from what would have been done if the spreadsheet had been correct.

Considerable research has been conducted into the prevalence, detection, and prevention of these spreadsheet errors. Estimates range from a majority of spreadsheets having errors of some type to virtually all spreadsheets having errors. One study even states that errors are not reduced with experience in building spreadsheets. Three of the best research centers (with extensive websites and links to research) are

- European Spreadsheet Risks Interest Group (www.eusprig.org)—in particular, it maintains a list of "horror stories" documenting the extensive costs associated with spreadsheet errors
- Spreadsheet Research (panko.shidler.hawaii.edu/SSR/index.htm)
- Spreadsheet Engineering Research Project (mba.tuck.dartmouth.edu/spreadsheet/)

These sites are worth visiting. They have links to research papers as well as best practices and tools to assist with identifying and preventing mistakes in spreadsheets. A number of tools are commercially available and Excel has some built-in tools for identifying spreadsheet errors. Trace Dependents and Trace Precedents (found on the Formula toolbar) can be used to identify which cells in a spreadsheet are linked to other cells. This can be particularly useful if used in conjunction with a model visualization, such as an influence chart. For example, Figure 1.9 shows the cell phone pricing spreadsheet with Trace Precedents set on the profit per customer cell. It shows that four cells are used in the formula for profit per customer, and the influence chart in Figure 1.2 shows that there are four arrows pointing to the calculation of profit per customer. The influence chart and the spreadsheet should match; links in the influence chart should correspond to the presence of a cell in a formula.

The Data Validation tools (found on the Data toolbar) permit you to restrict the types of values (e.g., numbers, dates, etc.) that can be used in particular cells as well as the range of acceptable values. Data Validation will circle cells with errors so that they can be further investigated.

Automated Error Checking will identify potential cells with errors, such as when the cell contains a formula that does not look like the

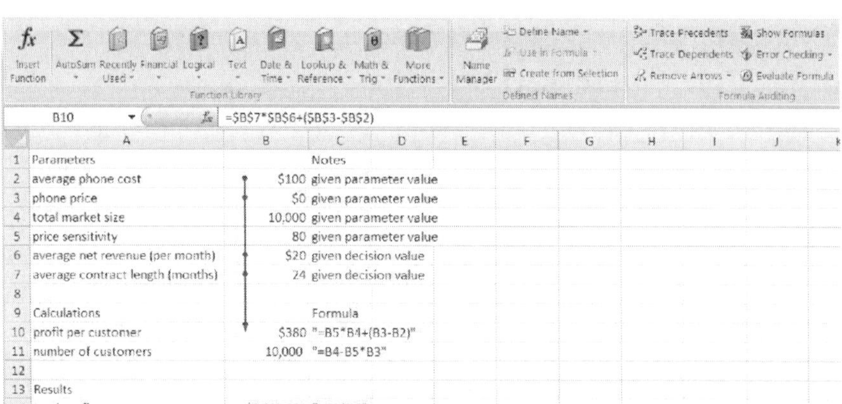

B10 =$B$7*$B$6+($B$3-$B$2)

	A	B	C
1	Parameters		Notes
2	average phone cost	$100	given parameter value
3	phone price	$0	given parameter value
4	total market size	10,000	given parameter value
5	price sensitivity	80	given parameter value
6	average net revenue (per month)	$20	given decision value
7	average contract length (months)	24	given decision value
8			
9	Calculations		Formula
10	profit per customer	$380	"=B5*B4+(B3-B2)"
11	number of customers	10,000	"=B4-B5*B3"
12			
13	Results		
14	total profit	$3,800,000	"=B9*B8"

FIGURE 1.9
Auditing tools.

formulas in adjacent cells. This can help locate the (all too common) copy and paste errors.

A number of commercial products are available that go beyond these simple tools. One of the best practices for building spreadsheets is to utilize such tools, but they can only identify certain types of errors.

A survey of the types of spreadsheet errors can be found in Panko and Aurigemma (2010).* They classify errors first into two types: culpable and blameless. There are methods for identifying and preventing culpable errors, but these are similar to means used to reduce other types of intentional errors, such as employee stealing, sabotage, etc. These are primarily organizational issues rather than technical ones, although some may be detected with the same tools that are used to detect blameless errors.

Blameless errors are divided into quantitative and qualitative errors, with quantitative errors further subdivided into planning errors and execution errors. These can be further delineated into a number of categories, such as copy and paste errors, overconfidence, archiving errors, etc. Readers should spend some time at the preceding sites to get a good sense of the frequency, types, and consequences associated with the myriad kinds of spreadsheet errors. One observation that is not generally emphasized is that there is no hierarchy to these errors. While we suspect that qualitative errors are in some ways "more serious" than quantitative errors, the consequences cannot be easily ranked. A single copy and paste error can have more practical significance than an important conceptual error, particularly in a complex spreadsheet. It is probably easier to detect the quantitative errors since they are more amenable to automated auditing tools, but this does not mean that the errors are less serious

1.6 Conclusion: Best Practices

At the outset, we emphasize that there is no "one size fits all" methodology to follow in spreadsheet modeling. Thus, we humbly submit the following "recommendations" rather than rules:

- Begin with a visualization of the problem to be modeled rather than a blank spreadsheet. Although more experienced users will often skip this step, faced with novel and complex problems, they will usually start here. If you cannot picture the logical structure of your spreadsheet, then it is not likely to emerge from a spreadsheet.

* Panko, R.R. and Aurigemma, S. 2010. Revising the Panko-Halverson Taxonomy of Spreadsheet Errors, *Decision Support Systems*, 49: 2, 235–244.

- Do not begin by entering data in a blank spreadsheet. Most problems have some input data (e.g., a set of prices, a list of characteristics, etc.) that can readily be placed in a spreadsheet. Novice modelers struggle for a place to begin and thus are tempted to enter these data, knowing that they will use them later on. However, the way in which the data are entered should be governed by the logical structure of the problem, rather than by the order in which the numbers are encountered. Once the data are entered in the spreadsheet, users are then forced to work around the (arbitrary) order in which they were entered. It is always better to have a logical structure in mind and then place the data in a form and format in which they belong.
- Make simplifying assumptions; try to pare the problem down to its essential core. Spreadsheets are natural environments for expanding your model (since copying cells is so easy), so it is typically best to start small and only expand when you have captured the essential relationships in your model.
- It may help to look first at available data to help guide the construction of your model, but do not worry too much about missing information. If a factor is important to your model, it should be in your model. It is always better to do additional research to find estimates for missing data that are critical than to leave them out of your model because they were not available to you initially. Alternatively, consider constructing the model differently so that unknown parameters may be avoided.
- Separate your model into sections: parameters, decisions, calculations, and outputs. Strive to have numbers (or distribution parameters for simulation) in the parameter and decision sections, and only formulas in the calculations and output/objectives sections.
- Document extensively. Insert comments or provide columns for comments that list the source for data and formulas as well as dates for when these are obtained. The more frequently a spreadsheet is to be used by other people, the more important this becomes since subsequent users of the spreadsheet are not likely to be familiar with the source material. Spreadsheets will outlast their originators, so this documentation will be lost if it only exists in the mind of the original creator of the spreadsheet.
- Test each formula as you enter it. Try extreme values to make sure it works as intended. Use appropriate auditing tools. When you are copying and pasting data, be especially careful to use the correct target range of cells. (An easy thing to check is whether the last cell you paste into has the correct formula in it.) Pay particular attention to relative and absolute cell addressing.

- Try to keep your formulas simple. If you find yourself entering a lengthy formula, try breaking it into several steps. Novice modelers impress themselves (or others) when they build a complex formula and it does what they want. While this may appear impressive, it is a typically a poor modeling habit. Spreadsheets have plenty of room and it is easier to troubleshoot your formulas if they are small than if they are large.
- If you find yourself with calculations columns without headers or values entered without labels, go back and find appropriate narrative labels for them. Commonly, unlabeled columns reflect that the modeler was not really sure what the formula was intended to accomplish. It will also be easier for other users of your spreadsheet if everything has a descriptive title or label to inform them of what its purpose is.
- Build graphs to illustrate the effects of different input data on your objective cells. This is an excellent way to communicate your results as well as a good way to make sure your model is working as intended. The graph should make intuitive sense. If it looks odd, then it may mean something is wrong with your model. For example, if your results show no sensitivity to a key parameter (or one that you thought was important), then quite possibly you have failed to use that parameter in the correct manner.
- Organizations should strive to establish quality control procedures for spreadsheet development. If you work in an organization, do you have established procedures? Most do not, but should. Revisit the "horror stories" on the EUSPRIG website if you need to be convinced.
- If you work on building models with others (and we recommend that you do), make sure you establish effective collaboration methods. It is not unusual to find two people making the same error in a spreadsheet if they work together. A good technique would be to work jointly on a visualization of a problem and then to create the spreadsheet prototype separately from your visualization. Comparison of models and results is usually informative.
- Protect against overconfidence! Spreadsheet modelers are prone to believe in their models, and they do so more strongly the more effort they expend in their creations. Research confirms these tendencies toward unwarranted overconfidence. Modelers also tend to be overconfident in the values they place in their spreadsheets. Often a value is chosen as a temporary placeholder, based on limited information, but it is subsequently perceived as more certain than it actually is. Possible protection against this is to use Monte Carlo simulation to represent the uncertainty about key parameter values. This is the subject of the next chapter.

TABLE 1.1

Health Insurance Plans

Plan	Monthly Cost	Individual Deductible	Out-of-Pocket Maximum	Copay for Inpatient Benefits	Office Visit Copay
Bronze	$347	$2000	$5000	20%	$30
Silver	$469	$500	$2000	15%	$20
Gold	$603	$0	Does not apply	0%	$20

CHAPTER 1
Exercises

For each of the following exercises, develop an appropriate visualization and then build a working spreadsheet to examine a base case. The base case should simplify the problem to its essential components and should not consider any uncertainty. (You will have plenty of opportunities to do this in the following chapters.)

1.1 HEALTH INSURANCE CHOICES

Health insurance decisions are important and complex. Many plans are available and they differ in numerous dimensions—virtually all are expensive. Your task is to build a model that will help individuals choose an appropriate plan from among a number of options. For this exercise, Table 1.1 provides data extracted for individual coverage.*

Assume that this individual's usage patterns are representative of the average in the population: three office visits/year/person and total health care services used, for which the percentage copayment applies: $7,000 per person. To simplify the way these plans work, assume that an insured individual is responsible for the deductible and then the copay until out-of-pocket maximum is reached, after which the insurance pays 100%. Which plan is cheapest for this person?

1.2 ELECTRIC VEHICLES

Consumers are becoming more sensitive to concerns about global warming, as well as concerns about their spending on gasoline. Electric hybrid vehicles offer an alternative that may help on both counts, but the initial purchase is more costly. Build a model to estimate the break-even additional cost for a hybrid vehicle under the following assumptions: The

* These prices and coverage features are stylized versions of information obtained from the Massachusetts Health Connector (www.mahealthconnector.org) on September 11, 2010, for a 35-year-old individual living in the 02145 zip code. The features approximate those provided by the choice of plans, but do not match any particular plan offer since the plans vary in many complex dimensions.

automobile is driven 11,000 miles per year, lasts for 12 years, and two types of engines are available:

- A conventional gasoline engine, which averages 37.7 miles per gallon
- A hybrid electric engine, which averages 52.7 mi/gal on gasoline and gets 4 mi/kwh on electricity (Half of the annual mileage will be made from each energy source.)

Examine the break-even extra cost (in terms of net present value) for the hybrid engine for gasoline prices of \$3.50/gal and electricity prices of \$0.15/kwh. Assume that the discount rate is 5%.

1.3 INVESTING IN EDUCATION

Choosing to go to college is one of the biggest lifetime decisions people make. Economists view this, at least partially, as investing in human capital. Ignoring the consumption value of college, there is certainly an expectation that income will be higher as a result of getting the degree. The data bear this out: Mean annual earnings for 25- to 29-year-olds with a high school degree are \$26,322; with a college degree, these increase to an average of \$43,163. Professional degree holders average \$55,389.* Income rises faster with education as well: In the 55- to 59-year-old age group (the peak earning age), high school graduates average \$36,215 of income, college graduates average \$65,180, and professional degree holders average \$138,075. Higher income levels will be subject to higher taxes. Use 4%, 8%, and 13% as the average tax rates that apply to high school graduates, college graduates, and professional degree holders, respectively.

Additional data concern the cost of college. Tuition and fees (per year) average \$5,950 at public colleges and \$21,588 at private colleges (65% of college students are enrolled at public colleges and 35% at private ones). Assume that college takes 4 years. Professional education averages \$30,000/year and requires an additional 2–4 years (use 3 years as an average in this exercise). For textbook expenses, use an average of \$750/year. We ignore the room and board costs because food and lodging are necessary regardless of whether one attends college, but do include the opportunity cost (lost earnings) of attending college rather than working full time.

A financial measure of investing in college (or professional schooling) is the net present value (NPV). Estimate the NPV of investing in a college degree (taking 4 years), assuming a discount rate of 7%. Also, estimate the NPV of investing in professional school (say, an MBA), taking an additional 3 years. Assume a person aged 20 years and a retirement age of 65 years.

---

* Income data come from the US Census Bureau, Current Population Survey, 2009. Tuition rates come from the National Center for Educational Statistics. Unemployment rates (used in Exercise 2.3 in Chapter 2) come from the Bureau of Labor Statistics.

TABLE 1.2
Customer Lifetime Value

Year	Retention Rate	Orders/Year	Average Order	Cost of Goods Sold	Acquisition/Marketing Costs
1	60%	1.8	$85	70%	$60
2	70%	2.5	$95	65%	$20
3	80%	3.2	$105	65%	$20

1.4 CUSTOMER LIFETIME VALUE

Customer Lifetime Value (LTV) is a concept developed to consider the fact that it generally costs something to acquire customers and that the profits rely on retaining them over time. Consider the hypothetical data in Table 1.2 concerning customer profitability over a 3-year time frame. Assume that your company has 100,000 initial customers. Estimate the LTV, which is defined as the cumulative net present value of the profits divided by the original number of customers. Use a discount rate of 15%.

1.5 NETWORK ECONOMICS

Communications technologies usually exhibit network effects: Their value increases more than proportionally with the number of users. For example, instant messaging services are more valuable the more people with whom you can exchange messages. Under such conditions, market growth can be self-sustaining, provided that enough users adopt the service. But, such markets must first exceed critical mass: the number of users required in order for the user base to grow. Failure to achieve critical mass may result in the user base unraveling. Figure 1.10 illustrates the situation.

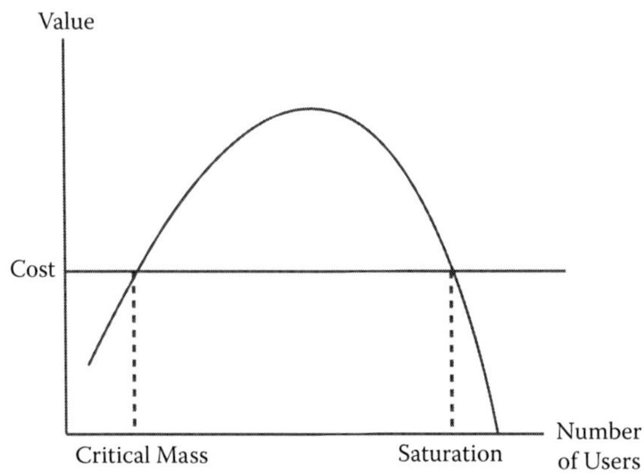

FIGURE 1.10
Network value function.

The curve in Figure 1.10 represents the value to the marginal user under the following conditions. Each user has an intrinsic value that represents the value to him or her, assuming that everybody else is also a user. Since people differ, these intrinsic values will differ across people. In this exercise, assume that the intrinsic values are given by $100(1 - F)$, where F is the fraction of the population that uses this service. This means intrinsic values start at \$100 (for the person with the highest valuation) and range down to \$0 for the person with the lowest value.

The actual realized value for each user is given by the intrinsic value multiplied by F: If everybody subscribes ($F = 1$), then each user gets his or her intrinsic value. If nobody subscribes, then nobody gets any value. If 50% subscribe, then everybody gets half of the intrinsic value. So, the actual realized value is given by $100F(1 - F)$, and this results in a curve such as what is shown in Figure 1.10.

Assume that the marginal cost is constant with the number of users and equals \$5. Figure 1.10 has three equilibrium points. To the left of the critical mass point, subscribers will have values less than the cost and will stop subscribing, eventually driving the equilibrium to zero. Critical mass represents an unstable equilibrium. As soon as the number of users rises above critical mass, values will exceed costs and the subscriber base will grow until the saturation level is reached. If the user base ever exceeds the saturation level, then there will be users with values less than the cost and they will eventually drop their subscriptions.

Determine the critical mass and saturation levels under these conditions.

1.6 PEAK LOAD PRICING

Electricity demand varies over time. It is expensive to maintain generating capacity sufficient for meeting the peak demand—especially if that capacity sits idle the rest of the time. In this exercise, assume that there are two hourly demand levels: peak and off-peak. Both are influenced by price. For the peak demand, assume that $Q = 100 - P$, while off-peak demand is given by $Q = 80 - 1.5\,P$. Peak demand is experienced 8 hours each day (for example, from 10 a.m. until 6 p.m.). Assume that capacity costs, expressed on a daily basis, equal 600 for each unit of capacity. (One unit of capacity can serve one unit of demand per hour.) There are no other costs to consider.

Compare the profitability of two pricing schemes: first, a single price of 50 for every hour of the day, and, second, a peak load pricing scheme that charges 75 during peak hours and 25 during off peak hours. Also, show the demand profile over a typical day under each of these pricing structures.

1.7 PROJECT MANAGEMENT

A new product development cycle involves a number of interrelated activities. Some activities can be pursued simultaneously, while some can only be undertaken after others are completed. Table 1.3 shows these stages for a typical new product. What is the expected total time to completion for this project?

TABLE 1.3

Project Activities List

Activity	Description	Required Predecessors	Duration (Months)
A	Product design	None	5
B	Market research	None	6
C	Production analysis	A	2
D	Product model	A	3
E	Sales brochure	A	2
F	Cost analysis	C	3
G	Product testing	D	4
H	Sales training	B, E	2
I	Pricing	F, H	1
J	Project report	F, G, I	1

1.8 HOLLYWOOD FINANCE

The worldwide movie industry is worth over $100 billion annually and the focus of a good portion of popular culture. Develop a model to estimate the profitability for an individual film, using the following information*:

- An average U.S. movie costs $72 million to produce and $37 million to promote. Assume that this takes place over a 6-month time period.
- Domestic box office receipts average $19.6 million and are earned during the first month of a film's release.
- Foreign box office receipts average about the same as domestic box office receipts, earned during the first 2 months of release.
- Home video rentals average 80% of the total domestic box office receipts and occur an average of 1 month after domestic theatre runs are finished.
- Home video sales average 220% of the domestic box office receipts and occur at the same time as the rental income.
- Premium pay TV revenues average $7 million/film, occurring over the 12-month period after video sales end.
- Network and free cable television revenues average $7 million/film, earned over the 12-month period following pay TV runs.
- Syndication averages around $2 million total, earned over the final 5-year period.

Develop a visualization and a spreadsheet to estimate the average profitability of a U.S. movie. Measure profitability by the internal rate of return.†

* These are stylized facts, based to some extent on data from Young, S. M., J. J. Gong, and W. A. Van der Stede. 2010. The business of making money with movies. *Strategic Finance* 91 (8).

† The internal rate of return is the discount rate that makes the NPV equal to zero. Use the Excel IRR function for this. (However, note the caveats about this function in Exercise 2.8 in Chapter 2.)

2

Basic Monte Carlo Simulation in Spreadsheets

LEARNING OBJECTIVES

- Understand the importance of uncertainty for model building and decision making.
- Learn about Monte Carlo simulation and how it is a good way to take into account risk and uncertainty in your models.
- See how to build Monte Carlo simulation in Excel spreadsheets.
- Acquire the basic skills of setting up a (Monte Carlo) simulation model and interpreting its results.

2.1 Introductory Case: Retirement Planning

Retirement planning can be greatly facilitated by using models. People need to make complex and interdependent decisions regarding how much to save for retirement, when to retire, and what lifestyle to enjoy after retirement. We developed a fairly simple model in Chapter 1 for a 30-year-old person beginning a new retirement account (though our model can readily be adapted to a number of different characteristics). This model, however, is only taking into account an "expected" case and is of limited use because it omits the reality that there are many uncertainties that are critical to making retirement plans. As the required disclaimer states, "Past performance is no guarantee of future results." We all know that investment returns are uncertain. Similarly, we know that health and employment are also uncertain and this should be accounted for in our planning.

We will use the retirement model from Chapter 1 to show how it can be enhanced to account for these critical uncertainties and how the results can increase our understanding of retirement decisions.

2.2 Risk and Uncertainty

There are many definitions of risk and they are often ambiguous. In his classic work, Knight (1921) distinguished between risk and uncertainty wherein the former is something measurable and the latter is not.* More recently, Taleb (2007) has popularized the notion of important risky events for which the probability of occurrence is difficult, if not impossible, to measure.† He argues that these are the truly important risks, but they are ones for which there is little or no evidence on which to base analysis. For our purposes, we will not distinguish among the various definitions and we will not confine ourselves to measurable risks.

The presence of uncertainty means that there are probabilities attached to different potential outcomes. We will sidestep for now the issue of how such probabilities and uncertainties may be estimated from data or estimated by experts and will just posit that probabilities can be attached to the various outcomes. A *probability distribution* is defined as the set of possible outcomes, each with its associated probability of occurring. There are two types of probability distributions: discrete distributions for outcomes that can only take on a discrete set of values and continuous distributions for outcomes that can be any value within a certain range. Figure 2.1 illustrates a discrete (the individual bars) and a continuous (the smooth curve) probability distribution.

With discrete distributions, all probabilities have to range between zero and one, and they must sum to one. For continuous distributions, the area under the curve always has to be one.

The amount of data available is exponentially increasing. One recent estimate of the amount of digital content is nearly 500 billion gigabytes—enough to form a stack of books stretching from the Earth to Pluto 10 times.‡ However, no matter how much data you have, the only thing we know for sure is that using them to forecast the future (without allowing for uncertainty) will likely give you the wrong answer! Whatever the past or present looks like, there is no guarantee that the future will behave the same; in fact, this is why decision makers have a marketable skill. It is precisely the fact that the future cannot be predicted with certainty that makes decisions difficult. No amount of data can change this fact. In the immortal words of Yogi Berra: "The future ain't what it used to be."§

When developing models to support decision makers, it would therefore be very useful to include our uncertainty quantitatively within the analysis. This will help in getting better insight into what can actually happen and not

* Knight, F. 1921. *Risk, Uncertainty and Profit*. Boston: Hart, Schaffner, and Marx.

† Taleb, N. 2007. *The Black Swan: The Impact of the Highly Improbable*. New York: Random House.

‡ Internet data heads for 500 bn gigabytes, guardian.co.uk, May 18, 2009.

§ Berra, L. P. 2002. *When You Come to a Fork in the Road, Take It!: Inspiration and Wisdom from One of Baseball's Greatest Heroes*. New York: Hyperion.

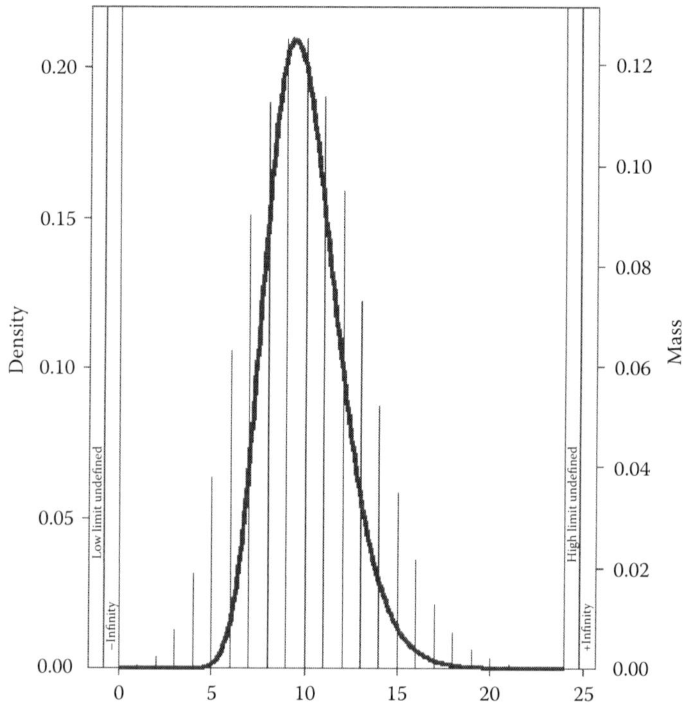

FIGURE 2.1
An example of discrete and continuous probability distributions.

just what we think is "most likely" to happen. The way we can do this within a (spreadsheet) model is that we will include uncertainty in the relevant inputs to our model. Such inputs may be based on historical data such as past stock returns or on expert judgments about the uncertainty about future stock returns, or (frequently) on combinations of both. To include our uncertainty about the future, we will need to replace some of the fixed parameters (e.g., our assumption about a rate of return of 7% per year in Chapter 1) with representations (probability distributions, as shown in Section 2.4) that take into account that more than one value is possible and we have uncertainty.

2.3 Scenario Manager

We have already seen in Chapter 1 that we can use "what if" analyses to examine the impact of different parameter values on our objectives. Excel contains a potentially useful tool to do such "what if" analysis in a semiautomatic way; it is called the Scenario Manager and is found on the Data ribbon under What If Analysis. To see how this tool works, let us return to our cell

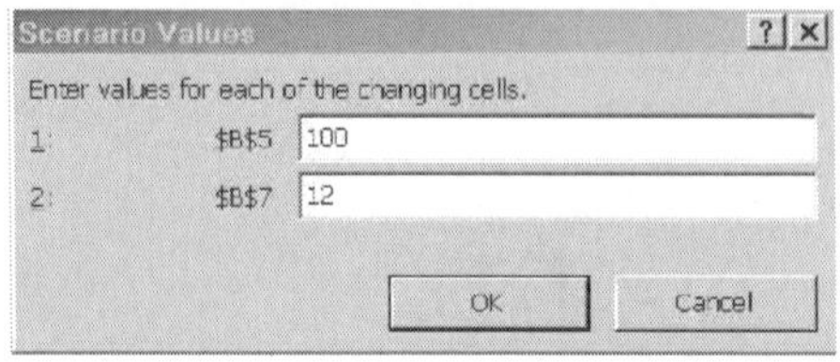

FIGURE 2.2
Scenario Manager setup window.

Scenario Summary	Current Values:	Pessimistic	Optimisic	Base Case
Changing Cells:				
$B$5	80	100	60	80
$B$7	24	12	30	24
Result Cells:				
$B$14	$3,800,000	$1,400,000	$5,000,000	$3,800,000

Notes: Current Values column represents values of changing cells at time Scenario Summary Report was created. Changing cells for each scenario are highlighted in gray.

FIGURE 2.3
Scenario Summary results window.

phone pricing model from Chapter 1. Figure 1.3 showed that there were two principal uncertain parameters in our model: the average contract length and the demand sensitivity. A common approach for modeling key uncertainties is to model pessimistic or optimistic scenarios (sometimes referred to as worst- and best-case scenarios). For example, we might believe that the pessimistic scenario would involve an average contract length of only 12 months (rather than the base case of 2 years) and a higher demand sensitivity of 100 (rather than the base case of 80). We can define this pessimistic scenario with the Scenario Manager. When you select Excel's Scenario Manager, you first assign a name to your scenario and list the cells that you wish to change in the scenario. Figure 2.2 shows the next dialog, where you assign the values to those cells that represent the chosen scenario.

We also defined an optimistic scenario, using the values of 30 months for the average contract length and 60 for the demand sensitivity, and a base case scenario using the original values. After defining the three scenarios, we click on Summary and choose B14 (total profit) as the result cell. The result is shown in Figure 2.3.

The summary shows us that our profits range from $1,400,000 to $5,000,000 with the base case being $3,800,000. The Scenario Summary is a concise way to evaluate these scenarios and is particularly useful when you need to evaluate a lot of separate scenarios.

However, we caution against using the Scenario Manager for analysis purposes.* In fact, the Scenario Manager has the danger of appearing to model

* When evaluating the outcomes of different discrete decisions, the Scenario Manager may be particularly useful.

the uncertainty without really modeling it. Just how likely are the pessimistic and optimistic scenarios? Are their probabilities equal? Who decided what "pessimistic" meant? Or, if it was defined as a worst-case scenario, does this mean that it is impossible to be worse or is the probability thought to be negligible (and then what does "negligible" mean)? We have seen many uses of the Scenario Manager in business and these questions are rarely answered.

In fact, if you try to use the Scenario Manager more carefully, it may be even less accurate and useful! Suppose that you obtained estimates for the pessimistic scenario by asking your marketing experts a precise question such as: For what contract length do you believe that there is only a 10% chance that it could be that short (or shorter)? Similarly, for the demand sensitivity, you ask for a value that is large enough that the probability of a larger value is only 10%. Then, if the pessimistic scenario uses these two values and if these two parameters are independent, the probability of the pessimistic scenario will actually be only 10% × 10% or 1%. Similarly, the optimistic scenario would only have a 1% chance of happening.

To obtain a scenario with only a 10% chance of occurring would require somewhat subtle calculations, and these would become quite complex if there are more than two parameters and some of them are dependent. As a result, the Scenario Summary *appears* to reflect uncertainty about the outcomes, but can often be misleading for decision-making purposes. Uncertainty can easily be overestimated or underestimated (usually the latter) using such a crude tool.

2.4 Monte Carlo Simulation

Monte Carlo simulation is a term coined by physicists working in the Los Alamos National Laboratory in the 1940s.* The name comes from the similarity to games of chance (e.g., blackjack outcomes or roulette wheel spins). While many implementations have been developed, they share the same general structure:

- Specify one or more probability distributions that describe the relevant uncertain variables in a problem.
- Create a random sample from each of these probability distributions.
- Compute an outcome associated with each random sample.
- Aggregate the results of a number of random samples to describe probabilistically the outcome of the decision problem.

* Metropolis, N. and S. Ulam. 1949. The Monte Carlo method. *Journal of the American Statistical Association* 44:335–341.

These four steps can be described in layman's terms: A Monte Carlo model is simulating not just the base case scenarios, but also any possible scenario through the use of probability distributions instead of fixed values. Monte Carlo then simulates the model thousands of times in order to get an understanding of the uncertainty around the possible outcomes.

Computer software is generally used to perform Monte Carlo simulation, but the idea can easily be imagined using paper, pencil, and a coin. Imagine that we are interested in estimating the probability of throwing two tails in a row in a coin toss. To do this, we could actually toss a coin many times. In other words, throw two dice, count the number of tails, and record its number. We do this many times (typically thousands) and then look at how many times out of the total tosses (e.g., 10,000 trials) we got two tails.* This will give us a very close approximation of the probability, which will be very close to 25% (e.g., we may get 0.2498 or 0.2509 as the fraction of the time we obtained two tails).

Fortunately, we do not have to do this "by hand," and we can rely on software to automate this procedure correctly. These software packages generate numbers from a uniform distribution, where every number between zero and one has an equal probability of occurring. Since all probability distributions entail cumulative probabilities that lie between zero and one, a random number from a uniform distribution can be used to represent the cumulative probability from such a distribution. Then, using a variety of techniques, it is possible to go from the randomly chosen cumulative probability to the underlying value associated with that cumulative probability.†

Figure 2.4 illustrates the procedure. The curve represents the cumulative probability distribution for a hypothetical probability distribution measured over some outcomes, x. The randomly chosen number from a uniform distribution gives a number, such as y_i, between zero and one, which can then be used to compute the value x_i that is associated with that particular number. The techniques for going from a uniform probability to the underlying random outcomes associated with it vary depending on each particular distribution.‡ Also, a variety of methods can be used to generate random numbers from a uniform distribution. Excel uses a particular algorithm but it is known to be not very robust. Fortunately, most simulation programs (including Monte Carlo Excel add-in packages) use well-tested algorithms

* Strictly speaking, Monte Carlo simulation becomes increasingly accurate as the number of times the procedure is repeated approaches infinity. In practice, we are typically satisfied after 10,000 iterations (in some situations, particularly when we are interested in very low probability events, more iterations may be required for precision).

† Do not worry if you did not fully understand exactly how software packages can generate random numbers. What is important is that software packages can generate random numbers from many different distributions and therefore greatly facilitate Monte Carlo simulation.

‡ A good example for the triangular distribution can be found in Evans, J. R. and D. L. Olson. 2002. *Introduction to Simulation and Risk Analysis*. Englewood Cliffs, NJ: Prentice Hall.

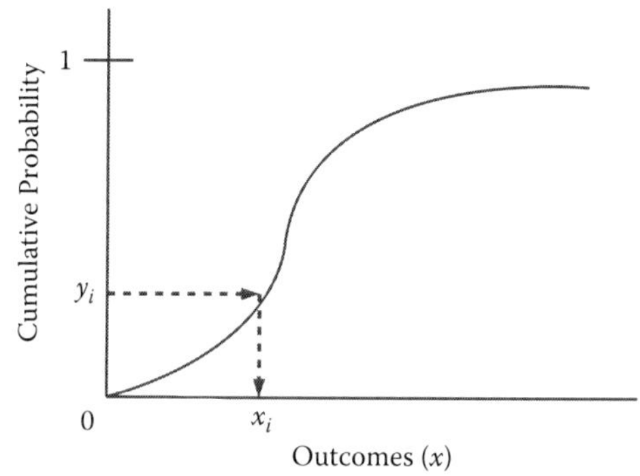

FIGURE 2.4
Cumulative probability functions and random number generation.

that ensure that each number between zero and one is equally probable to be chosen and that the sequence of generated numbers is random.*

For example, let us review the cell phone pricing example from Chapter 1. We may imagine that the average contract length is unknown, but we believe it ranges between 12 and 30 months, with 24 months being the most likely outcome. We might also believe that the demand sensitivity parameter ranges from 60 to 100, with 80 being the most likely outcome. We will assume that these two uncertain variables are independent (exploring the case where they are correlated in Chapter 7) and that both the contract length probability distribution and the demand sensitivity distribution are distributed according to a PERT distribution. (This is a distribution frequently used for modeling expert opinion.) It has three input parameters: the minimum, maximum, and most likely values. (We explore how distributions may be chosen in Chapter 4.)

These two PERT distributions are illustrated graphically in Figure 2.5. As you can see, the "peaks" of the two distributions are at the most likely values: 24 and 80, respectively. Further, the min and max values are where the distribution graphs touch zero, illustrating that we assume that the values cannot be lower than the min or higher than the max.

Conducting the simulation then depends on taking random samples simultaneously from these two distributions, computing the outcome of our cell phone pricing model, repeating this thousands of times, and aggregating the

* An easy way to think of random number generator is to take an eight-digit number (with a decimal point in front) and select the middle four digits (the random seed). Then, square this number; the result is another eight-digit number. The procedure is continually repeated with the resulting numbers being randomly distributed between zero and one. While this works most of the time, problems can emerge. For example, if the middle four digits produce .0001, then this procedure will get stuck at .0001, which is clearly not a random sequence of numbers. This algorithm would not be *robust*. Random number generators, such as that used in the ModelRisk software, have been tested and documented for robustness to avoid pitfalls such as these.

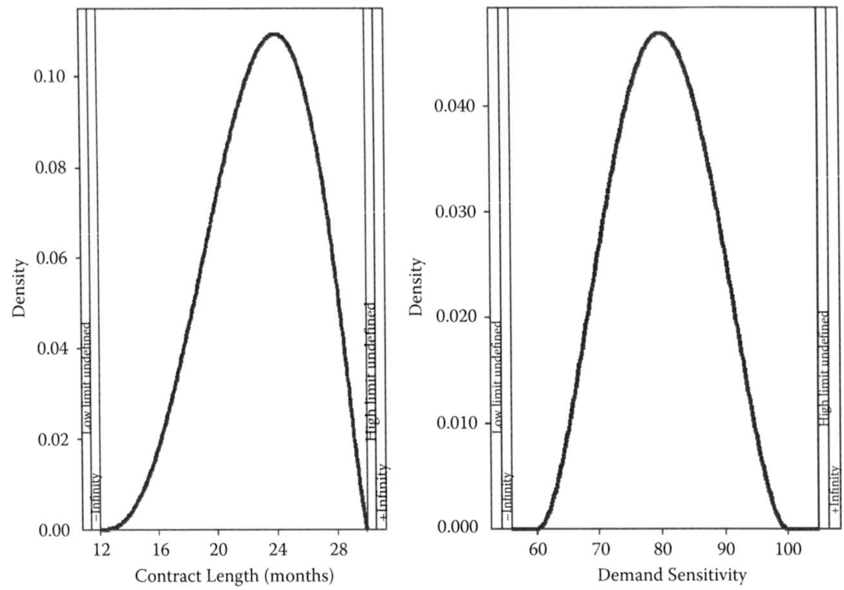

FIGURE 2.5
Probability distributions for the cell phone pricing model.

results. To accomplish this in an automated way, we turn to the Monte Carlo software packages, such as ModelRisk.*

2.5 Monte Carlo Simulation Using ModelRisk

In order to follow this example, we assume that you have installed the ModelRisk software. We are going to replace our two uncertain parameters (average contract length and demand sensitivity) with the probability distributions shown in Figure 2.5. The ModelRisk ribbon is shown in Figure 2.6, and we will use the Select Distribution button to enter our distributions.

Selecting the first option in the ModelRisk ribbon, called Select Distribution, presents us with a list of different sets of distributions. (We will discuss these in much more detail in Chapter 4.) Scrolling through the All Univariate distributions to the PERT and selecting it gives the dialog shown in Figure 2.7.

Regardless of the distribution type, the ModelRisk distribution window contains a number of consistent features. Along the top bar, the user can choose from different views of the probability distribution that the software

* See Appendix A for an overview of a number of alternative Monte Carlo Excel add-in software packages.

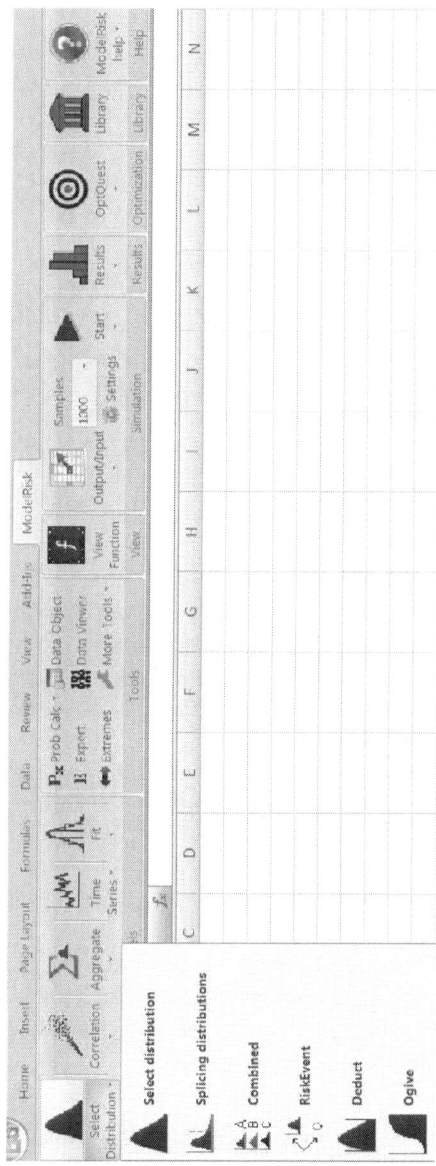

FIGURE 2.6
Select Distribution on the ModelRisk menu ribbon.

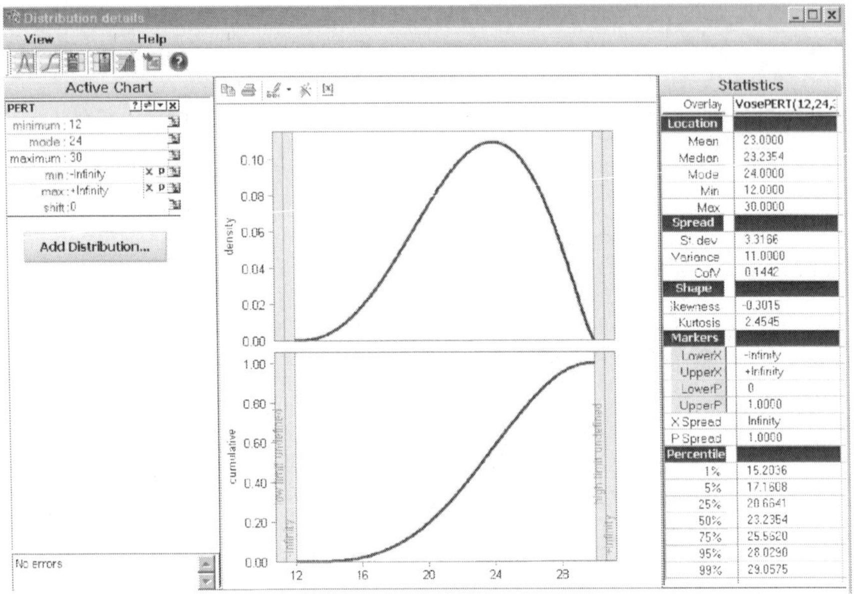

FIGURE 2.7
Select Distribution dialog box.

can show. The default (shown) has two graphs: One is for the probability density* and the second (below it) is the cumulative probability distribution. The former shows the probability density of each potential value for the contract length and the latter shows the cumulative probability of obtaining a value less than or equal to each of these values. To the right are a number of statistics corresponding to the distributions. The parameters for our distribution (in this case the minimum, mode or most likely, and maximum) are entered on the left side (where we can also restrict the range of values or shift the distribution). Note that these values can be placed on the spreadsheet in cells and the dialog can be linked to these cells—a much better modeling technique, since it really keeps the parameters visible on the spreadsheet and amenable to being changed easily by the user.

When we are done entering the parameters for our contract length distribution, we click on the Excel symbol button (which says "insert to sheet" when you hover over it) and insert it into the spreadsheet. For now, we will insert it into cell B7 in Figure 1.1 to replace our previously assumed (certain)

* The probability density of a continuous distribution is a somewhat difficult concept since it is not the same as a probability. However, similar to a probability, the higher the probability density is, the higher is the likelihood of a value within that corresponding region. The difference is that when we have a continuous distribution, we can only talk about the probability of being *between* two values, or above or below a certain value, and *not* about the probability of exactly a certain value. The probability density values themselves are therefore typically not of great interest.

Parameters	
average phone cost	100
phone price	0
total market size	10000
price sensitivity	=VosePERT(60,80,100)
average net revenue (per month)	20
average contract length (months)	=VosePERT(12,24,30)
discount rate	0.1
Calculations	
profit per customer	=-PV(B8/12,B7,B6)+($B$3-$B$2)
number of customers	=B4-B5*B3
Results	
total profit	=VoseOutput("Total Profit","$")+B12*B11

FIGURE 2.8
Revised cell phone pricing model.

contract length of 24 months. The Insert to Sheet button presents us with a number of options: We can paste the Distribution, other variants on the Distribution (such as the cumulative distribution, the inverse of the function, etc.), or an Object (covered in Chapter 3). In this case, we will paste the distribution here. We then perform the same steps for the price sensitivity parameter in cell B5 of Figure 1.1. After pasting the distribution into cell B5 and slightly adjusting our model to reflect the time value of money, our spreadsheet formulas are shown in Figure 2.8.

The discount rate appears in cell B8 and the profit per customer formula uses the PV formula, which calculates the present value of a series of payments.* In Figure 2.8, the two cells with our uncertain parameters now show two Excel functions in their place. Both are functions that start with the word "Vose." The syntax for ModelRisk functions is that they are Excel functions that begin with the name Vose, followed by what type of function they are, with a number of function arguments. In this case, the functions are PERT distribution functions, and the arguments are the parameters for these distributions that we have assumed.

Note a few important comments and good practices about these functions:

- They are Excel functions and can be entered and manipulated like other Excel functions. For example, instead of the PERT parameters of a minimum = 60, mode= 80, and maximum = 100, we could (and should) use cell addresses that contain these values.

* The PV formula has three required inputs: the discount rate (which we divide by 12 to get a monthly rate), the number of periods (which is our simulated contract length), and the monthly payment (given in cell B6). The negative sign is required because the PV formula returns a negative number as its output (representing the present value of a payment).

- As you gain familiarity with these ModelRisk distribution functions, you may bypass the Select Distribution dialog and directly enter the ModelRisk function and its arguments into the spreadsheet, just like any other Excel function. These functions are fully supported through help like all other Excel functions.
- These functions can be copied and pasted just like any other Excel function.
- If you want to see the Select Distribution dialog for any reason after you have entered it, simply click the View Function button, shown in Figure 2.9, on the ModelRisk ribbon and it will immediately take you to that dialog box.
- When the spreadsheet values (not the formulas) are displayed, you will notice that the entries in cells B5 and B7 keep changing every time you recalculate the spreadsheet and will range somewhere within the chosen distributions.

When you are viewing the spreadsheet without displaying the formulas, if you press the F9 key (the recalculate key), you will see the values in cells B5 and B7 keep changing. These are, in fact, single Monte Carlo simulation runs (also known as samples, single iterations, or trials), where ModelRisk is randomly selecting from each of the probability distributions we have entered.

To complete our first simulation, we must inform ModelRisk the outcome(s) in which we are interested. For now, let us focus on just the total profit (cell B14). Select cell B14 and click on the Output/Input button on the ModelRisk ribbon. Choose Mark Output/Input and you will get the dialog shown in Figure 2.10, where we have named our output Total Profit and indicated that it will be measured in dollars ($). (Note that these entries can also be cell addresses and entered by clicking on the appropriate cells.)

We are ready to run our Monte Carlo simulations. Click on the Settings button on the ModelRisk toolbar. Indicate that we will run one Simulation, take 10,000 samples, and use a manual random seed of zero.* This dialog box is shown in Figure 2.11.

The simulation is run by clicking the Start button on the ModelRisk ribbon. After the 10,000 simulations are complete, the results shown in Figure 2.12 should appear.

* The setting of a manual seed is not necessary and we usually do not do this when running simulations. The results shown in the book are based on the random seed value of zero, so if you want to reproduce the exact results shown in the book, you must set the random seed manually to zero. If you leave it on automatic, ModelRisk will randomly choose the random seed and you will get slightly different numerical results than those in the book (and different every time you run the simulation). Such variability is the norm with Monte Carlo simulation, since every 10,000 random samples will differ from any other 10,000 random samples unless you force them to be the same random samples each time. That is what the manual seed accomplishes. As discussed previously, when simulating a large number of iterations, the differences between runs become smaller and smaller.

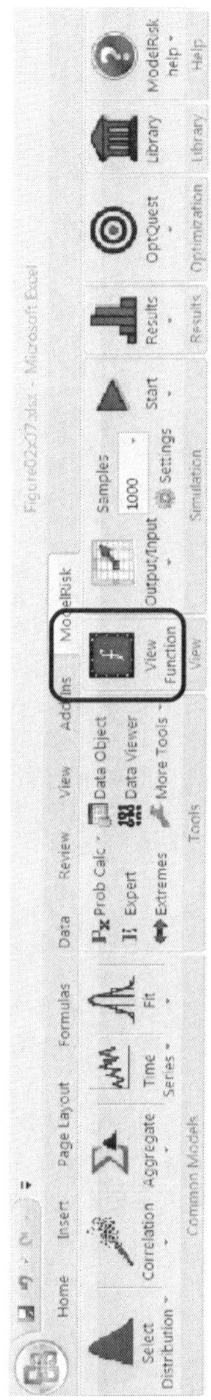

FIGURE 2.9
View Function button.

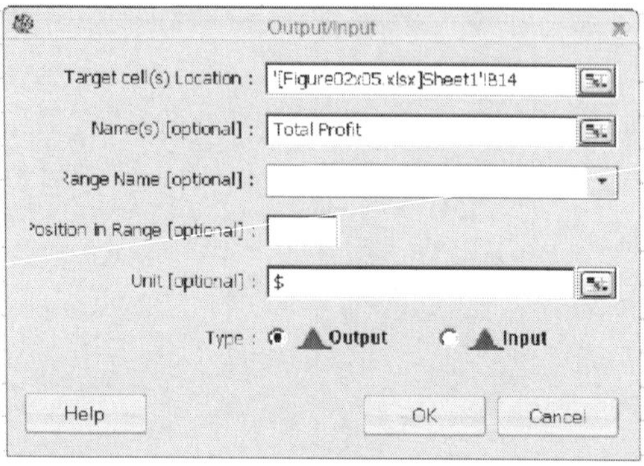

FIGURE 2.10
Mark Output/Input dialog box.

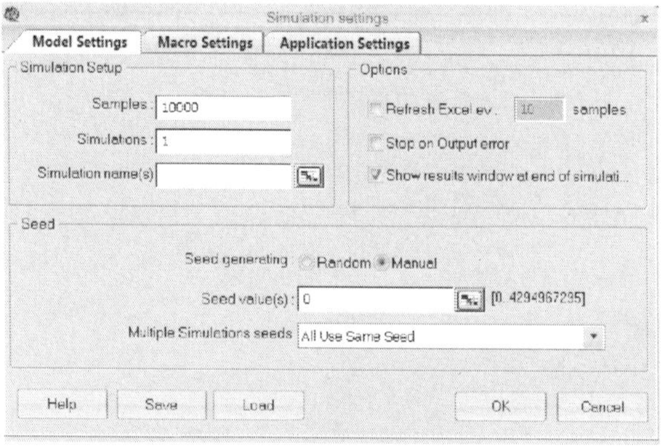

FIGURE 2.11
Simulation Settings dialog box.

The default view is a histogram of the results with sliders showing an 80% interval around the mean of the total profit distribution (ranging from approximately $2.4 million to $3.9 million).* A number of different views of these results can be selected by using the various display types shown in the Results window.† In particular, clicking on the Stats button will show a number of statistics regarding the simulation, such as the mean = $3,164,428, the St. dev. (standard deviation) = $548,721, and various percentiles of the distribution, such as the 10th percentile

* You can change the 80% interval by either dragging the sliders or using the define slider button just above the histogram and entering appropriate values. Note that you can enter either percentages or values; the use of values will have the display show you the percent of the simulations that fall above or below that value.

† Your Results window will show no inputs on the left side of the screen and this makes the Tornado chart (one of the display types) unavailable. We will explain and use inputs in our next example.

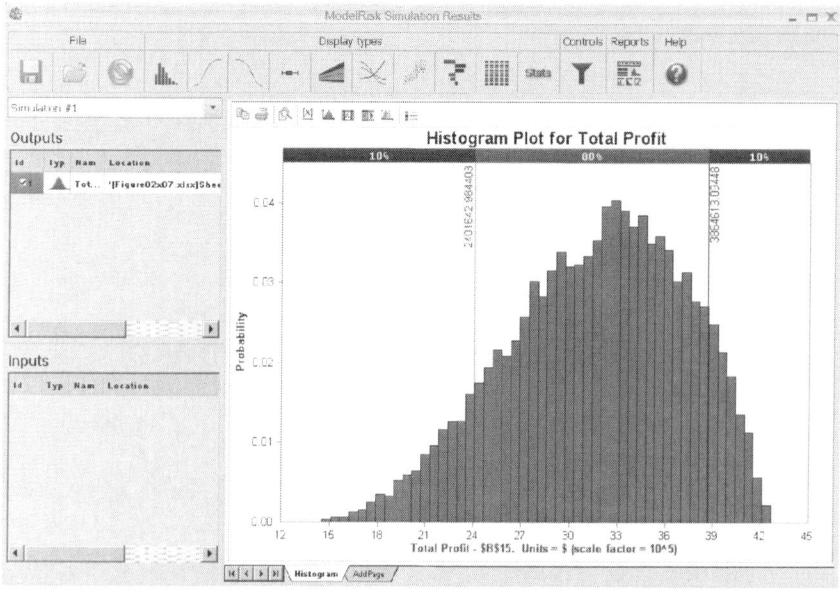

FIGURE 2.12
Initial simulation results.

($2,402,400—meaning there is a 10% chance of profits being less than this) or the 90th percentile ($3,865,313, for which there is only a 10% chance of profits being greater).

Astute readers may have noticed that, given the current model inputs, the price sensitivity parameter really has no impact on the results since we have assumed a handset price of zero (full subsidy). To explore different prices and the potential impacts of price sensitivity, we can modify our spreadsheet model to add a second column that calculates the profits for a second handset price—for example, $20. Figure 2.13 shows our adapted model, along with a second designated ModelRisk output cell (C15).

We run this simulation (with a random seed = 0.0) and obtain the results shown in Figure 2.14 (after checking both outputs in the left side of the box; this creates an overlay display of the two distributions).

The profits appear to be greater with a full subsidy than with an 80% subsidy. However, closer inspection (easily seen if you switch to the Stats view and look at the standard deviations) reveals that there is also more uncertainty about the outcome with the full subsidy. A good view of this is obtained by selecting the Ascending Cumulative Plot view of the simulation results, shown in Figure 2.15.

The cumulative distribution for the full subsidy lies to the right of that for the 80% subsidy, showing the higher expected profits. The greater variability is represented by the flatter slope of the cumulative probability function. (It takes a greater range of profit values to produce the same range of probable outcomes.)

	A	B	C
1	Parameters		
2	average phone cost	100	
3	phone price	0	20
4	total market size	10000	
5	price sensitivity	=VosePERT(60,80,100)	
6	average net revenue (per month)	20	
7	average contract length (months)	=VosePERT(12,24,30)	
8	discount rate	0.1	
9			
10	Calculations		
11	profit per customer	=-PV(B8/12,B7,B6)+($B$3-$B$2)	=-PV(B8/12,B7,B6)+(C3-B2)
12	number of customers	=B4-B5*B3	=B4-B5*C3
13			
14	Results		
15	total profit	=VoseOutput("Total Profit at Full Subsidy","$")+B12*B11	=VoseOutput("Total Profit at 80% Subsidy","$")+C12*C11

FIGURE 2.13
Cell phone model for two different prices.

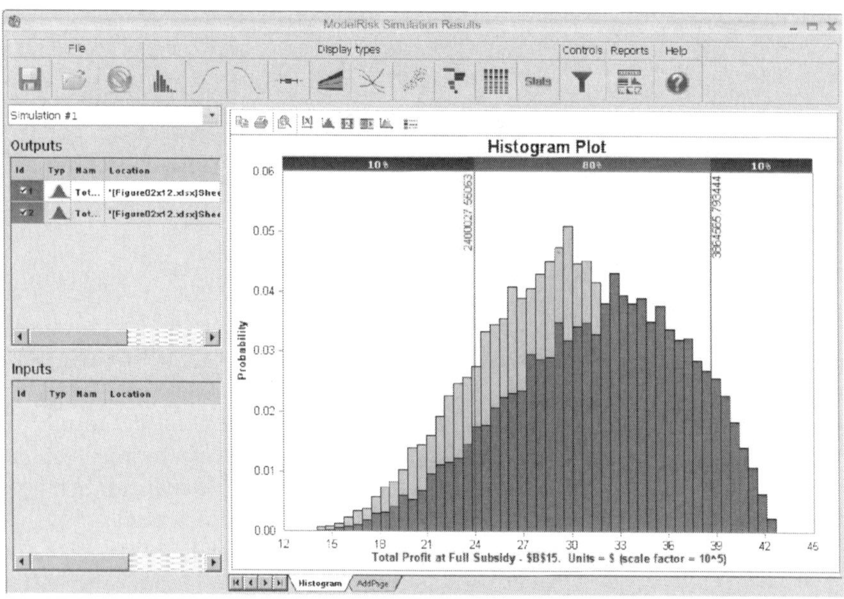

FIGURE 2.14
Overlay results.

This is a common insight provided by a risk analysis: There is a trade-off between risk and return. Expected profits are higher (by about $330,000) when we fully subsidize the handset than if we only offer a partial subsidy, but we are exposed to more risk. (The standard deviation rises by around $86,000.)*

We will further explore the use of Monte Carlo simulation by returning to our retirement example from the beginning of this chapter.

* In Chapter 7, we will discuss stochastic optimization, which could be used in this situation to examine whether there may be a "best" price to charge customers.

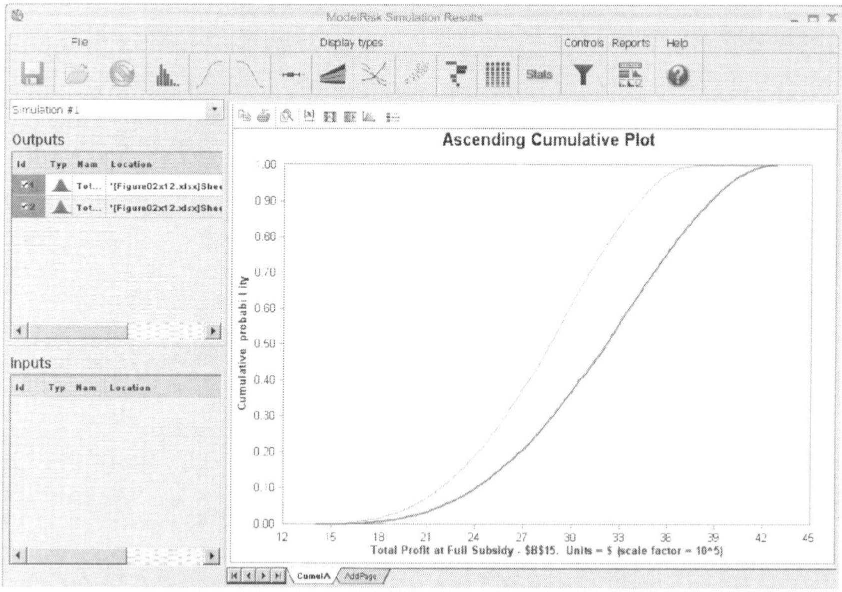

FIGURE 2.15
Cumulative distribution overlay.

2.6 Monte Carlo Simulation for Retirement Planning

Figure 2.16 shows our visualization of the retirement planning problem, with three key uncertainties represented: the average annual raise, the average annual living expenses (after retirement), and the annual return for the retirement fund.

Historical stock market data show that the annual real return of the S&P 500 over the 1950–2010 period was approximately normally distributed with a mean of 4.7% and a standard deviation of 16.4%.* We will also assume that the annual raise is uncertain, and we will use a PERT distribution with a minimum of 0%, a most likely value (mode) of 1%, and a maximum of 3%. Our uncertainty about future average postretirement living expenses will be assumed to follow a PERT distribution with a minimum of $40,000, mode of $50,000, and maximum of $70,000. This uncertainty is supposed to reflect uncertainty about what our individual's future lifestyle might look like.

* In Chapter 4, we will show how this can be derived from historical data and how to find the appropriate distribution to use. A normal distribution does not actually fit the data best; the best fitting distribution has longer tails—an important consideration for retirement planning since this means that extreme outcomes are more probable than the normal distribution permits. For now, we will use the normal distribution because it is probably more familiar to most readers.

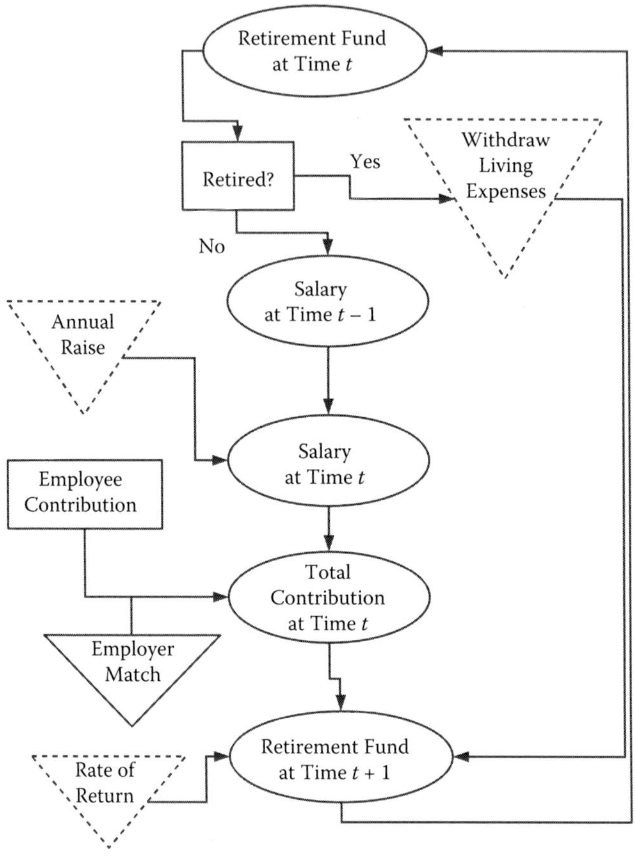

FIGURE 2.16
Visualization for retirement planning under uncertainty.

For example, one reason for the potentially larger annual expenses may be that health status declines and the individual incurs increased out-of-pocket health care costs. Finally, we assume that the retirement age is expected to be 65.*

We show these assumptions in Figure 2.17, but we must make an additional modification to our spreadsheet.† The annual return on the retirement fund is uncertain, but we would expect each year's return to be

* Further enhancements to the model would likely explore the retirement age more fully. It is partially a decision, partially random, and partially an objective. For instance, a person might want to minimize the retirement age subject to constraints on the size of the retirement fund (an objective). Alternatively, someone might be choosing the retirement age (a decision) in order to achieve particular goals of retirement lifestyle and length. To some extent, retirement age may be beyond a person's control (uncertain) because it may be impacted by health or by obsolescence of workplace skills.

† We will see in Chapter 3 that this additional modification is unnecessary if we use Distribution Objects in our model.

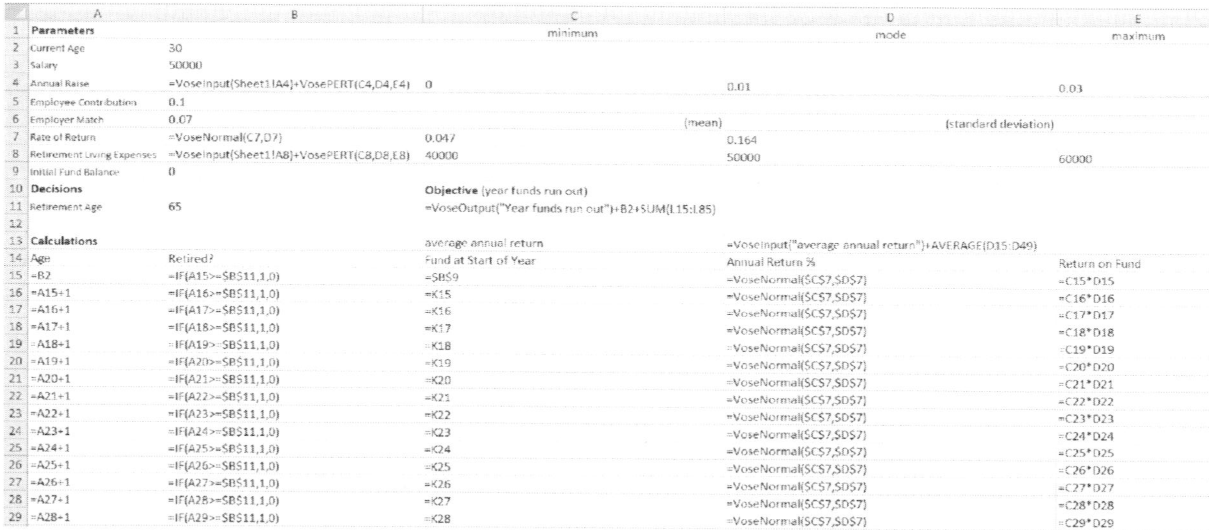

	A	B	C	D	E
1	**Parameters**		minimum	mode	maximum
2	Current Age	30			
3	Salary	50000			
4	Annual Raise	=VoseInput(Sheet1!A4)+VosePERT(C4,D4,E4)	0	0.01	0.03
5	Employee Contribution	0.1			
6	Employer Match	0.07	(mean)	(standard deviation)	
7	Rate of Return	=VoseNormal(C7,D7)	0.047	0.164	
8	Retirement Living Expenses	=VoseInput(Sheet1!A8)+VosePERT(C8,D8,E8)	40000	50000	60000
9	Initial Fund Balance	0			
10	**Decisions**		**Objective** (year funds run out)		
11	Retirement Age	65	=VoseOutput("Year funds run out")+B2+SUM(L15:L85)		
12					
13	**Calculations**		average annual return	=VoseInput("average annual return")+AVERAGE(D15:D49)	
14	Age	Retired?	Fund at Start of Year	Annual Return %	Return on Fund
15	=B2	=IF(A15>=$B$11,1,0)	=$B$9	=VoseNormal($C$7,$D$7)	=C15*D15
16	=A15+1	=IF(A16>=$B$11,1,0)	=K15	=VoseNormal($C$7,$D$7)	=C16*D16
17	=A16+1	=IF(A17>=$B$11,1,0)	=K16	=VoseNormal($C$7,$D$7)	=C17*D17
18	=A17+1	=IF(A18>=$B$11,1,0)	=K17	=VoseNormal($C$7,$D$7)	=C18*D18
19	=A18+1	=IF(A19>=$B$11,1,0)	=K18	=VoseNormal($C$7,$D$7)	=C19*D19
20	=A19+1	=IF(A20>=$B$11,1,0)	=K19	=VoseNormal($C$7,$D$7)	=C20*D20
21	=A20+1	=IF(A21>=$B$11,1,0)	=K20	=VoseNormal($C$7,$D$7)	=C21*D21
22	=A21+1	=IF(A22>=$B$11,1,0)	=K21	=VoseNormal($C$7,$D$7)	=C22*D22
23	=A22+1	=IF(A23>=$B$11,1,0)	=K22	=VoseNormal($C$7,$D$7)	=C23*D23
24	=A23+1	=IF(A24>=$B$11,1,0)	=K23	=VoseNormal($C$7,$D$7)	=C24*D24
25	=A24+1	=IF(A25>=$B$11,1,0)	=K24	=VoseNormal($C$7,$D$7)	=C25*D25
26	=A25+1	=IF(A26>=$B$11,1,0)	=K25	=VoseNormal($C$7,$D$7)	=C26*D26
27	=A26+1	=IF(A27>=$B$11,1,0)	=K26	=VoseNormal($C$7,$D$7)	=C27*D27
28	=A27+1	=IF(A28>=$B$11,1,0)	=K27	=VoseNormal($C$7,$D$7)	=C28*D28
29	=A28+1	=IF(A29>=$B$11,1,0)	=K28	=VoseNormal($C$7,$D$7)	=C29*D29

FIGURE 2.17
Spreadsheet model for retirement planning under uncertainty.

independent of the other years. So, we cannot just simulate 1 year's return and use that in our model for each year; instead, we must simulate each year's return. We do this by modifying column D of the spreadsheet to calculate the return by multiplying the fund size in column C of the spreadsheet by a function, VoseNormal (4.5%, 16.4%). Every year will now have its own simulated asset return.

Figure 2.17 shows the distributions we have entered along with the modifications to our previous retirement model. The annual rate of return is simulated in column D and the arithmetic average of these returns (during the working life) is calculated in cell D13. Note that the probability distributions refer to cell addresses holding the parameters of these distributions.

Note that we have also defined several inputs in our model (use the Output/Input button); we change the selection to Input and provide appropriate names to accomplish this. We have marked the average living expenses, average annual raise, and average annual return (cell D13) as inputs. Marking these inputs will allow us to use a Tornado chart to explore the relative importance of our various uncertain factors.

We are now prepared to run the simulation. Again, if you want to match our exact numerical outputs, set the manual seed to zero. We run 10,000 samples in order to get our results. Our results window is shown in Figure 2.18.

Immediately noticeable in Figure 2.18 is the tall bar at age 100. We built our model to go only to age 100, and in around 27% of the simulations the retirement fund was not exhausted. This is certainly encouraging for our retirement plans, but the fact that 47% of the time the fund does not last until our life expectancy of 81 years is worrisome (define the slider position to the left-hand value at 81 to see the 47% in the results histogram). Further insight is available through the Tornado chart. Click on the Tornado chart display type at the top of the Simulation Results window, and make sure all of the inputs are checked, to get Figure 2.19.

The Tornado chart shows the relative importance of our uncertain parameters. The most significant factor is the average annual return at age 60, followed (distantly) by the average annual raise and then by the retirement average living expenses.*

Our retirement plans appear to have a fair amount of risk, probably too risky for many people. What can we do about this? We possibly need to save more. We would like to compare the results for different decisions about what fraction of our salary to save each year. (The employee contribution was set at 10%.) Let us see what happens as we vary the employee contribution between 5% and 15%. To do this, we make use of the SimTable functionality in ModelRisk.

* The Tornado chart with a bar on the left-hand side shows a negative correlation, which indicates that higher values for this variable—average living expense—are associated with lower values for the objective—age when funds run out; there is a positive relationship for the other uncertain parameters.

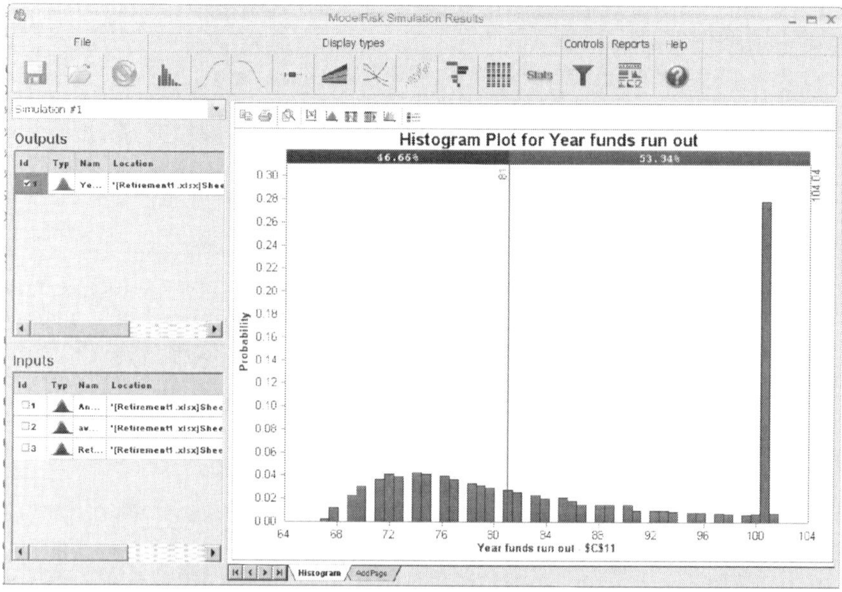

FIGURE 2.18
Results for the retirement model.

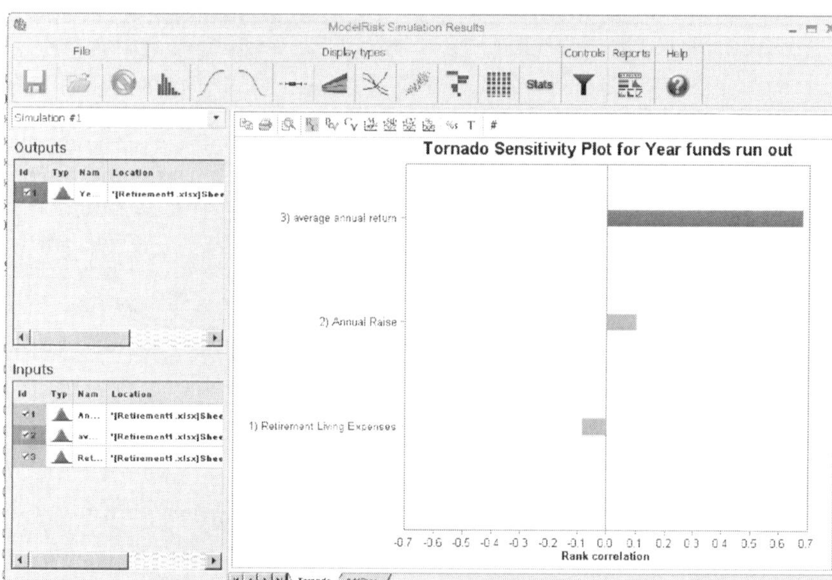

FIGURE 2.19
Tornado chart for the retirement model.

To set up a SimTable, use the following steps:

- On a clear part of the spreadsheet, place a column showing Simulation Number, with numbers running from 1 through the last simulation required. We are going to look at contribution percentages ranging from 5% to 15% in 1% increments, so this involves 11 separate simulations.
- In the next column, list the Contribution percentages (5% to 15%).
- Add as many columns of output parameters as you would like to calculate. In other words, in this example, we are interested in the year that funds will run out (cell C11), and we will be interested in the median year and the probability that the funds will run out before our expected lifetime of 81 years.* The function VoseSimPercentile, with 50% as the percentile, provides the median outputs. (The first argument of the VoseSimPercentile function refers to the output cell to use, the second refers to the percentile in which you are interested, and the third must refer to the cell you created that shows the simulation number.) The function VoseSimProbability provides the probability of being less than or equal to a designated value. When you first enter these functions, your spreadsheet should show "No simulation results" because the function can be evaluated only after a simulation is run.†
- Now you must tell ModelRisk where to use these varying Contribution factors. We want to place them in cell B5, so we enter the function VoseSimTable in that cell (i.e., we replace the formula that was previously in that cell). The first argument is the range of cells where we placed our range of values for the contribution fraction (g2:g12). The second argument is not necessary, but you must include two commas to bypass it. (Alternatively, you can insert the address for the range of cells that identify the simulation number, but this is optional.) The third argument is the value you want placed in the cell when you are not running a simulation. (We will use the original 10%.) The different arguments are explained in detail in ModelRisk's help.
- The final step is to click on the Settings and change the number of simulations from 1 to 11, as shown in Figure 2.20.

* The median is more relevant than the mean since we have truncated the spreadsheet at age 100. The median will not be sensitive to whether a particular run shows 100 years when it might show 120 years if we expanded the spreadsheet.

† These functions, based on the results of simulations, can be inserted anywhere in a spreadsheet (not only in SimTables). So, if you want to see the mean, 10th or 90th percentile, or other characteristics of an output cell, insert these functions in your spreadsheet. They will show "No simulation results" until a simulation is run; only after you complete running a Monte Carlo simulation will they show the appropriate results. If you base formulas on these cells (such as 1-VoseSimMean), then Excel will display an error until the simulation is run.

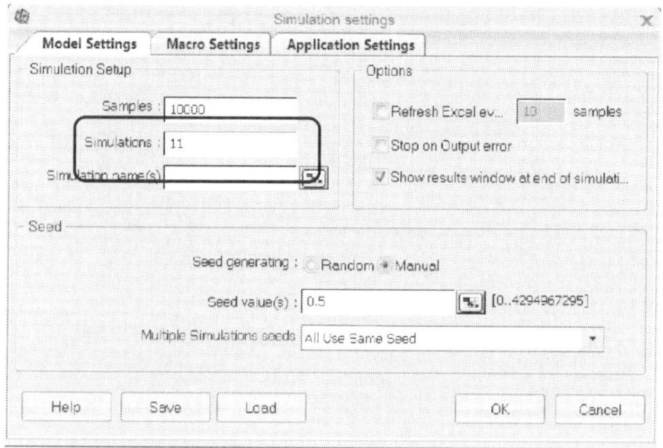

FIGURE 2.20
Simulation settings for the SimTable.

F	G	H	I
Simulation #	Contribution %	Median Age	Prob < = 81
1	0.05	=VoseSimPercentile($C$11,50%,F2)	=VoseSimProbability($C$11,81,F2)
2	0.06	=VoseSimPercentile($C$11,50%,F3)	=VoseSimProbability($C$11,81,F3)
3	0.07	=VoseSimPercentile($C$11,50%,F4)	=VoseSimProbability($C$11,81,F4)
4	0.08	=VoseSimPercentile($C$11,50%,F5)	=VoseSimProbability($C$11,81,F5)
5	0.09	=VoseSimPercentile($C$11,50%,F6)	=VoseSimProbability($C$11,81,F6)
6	0.1	=VoseSimPercentile($C$11,50%,F7)	=VoseSimProbability($C$11,81,F7)
7	0.11	=VoseSimPercentile($C$11,50%,F8)	=VoseSimProbability($C$11,81,F8)
8	0.12	=VoseSimPercentile($C$11,50%,F9)	=VoseSimProbability($C$11,81,F9)
9	0.13	=VoseSimPercentile($C$11,50%,F10)	=VoseSimProbability($C$11,81,F10)
10	0.14	=VoseSimPercentile($C$11,50%,F11)	=VoseSimProbability($C$11,81,F11)
11	0.15	=VoseSimPercentile($C$11,50%,F12)	=VoseSimProbability($C$11,81,F12)

FIGURE 2.21
SimTable setup.

Figure 2.21 shows the part of the spreadsheet we have created for the SimTable.

Cell B5 should now show =VoseSimTable(G2:G12,,10%). We run 11 simulations of 10,000 samples each and obtain the results shown in Figure 2.22.

The median year (the median is another word for the P50, the value of which we have a 50:50 chance to be above or below) that our funds run out rises continuously as we increase our contribution percent, and the probability of running out by age 81 drops. For example, increasing our contribution from 10% of salary to 15% increases the median from 83 to 91 years and decreases the probability we will run out by age 81 from 44% to 30%. These results can help with retirement planning by exploring how our risks change as we change our decisions. If we still find these results too risky, then we would need to explore further changes, such as planning to reduce our postretirement expenses, raise our retirement age, or even raise our savings rate further.

Simulation #	Contribution %	Median Age	Prob < = 81
1	5%	74	72.56%
2	6%	76	63.42%
3	7%	79	55.05%
4	8%	80	51.23%
5	9%	81	47.46%
6	10%	83	43.91%
7	11%	84	40.79%
8	12%	86	37.87%
9	13%	88	35.05%
10	14%	89	32.50%
11	15%	91	30.05%

FIGURE 2.22
SimTable results.

Industry Spotlight*

Denali Alaskan Federal Credit Union is a federally chartered financial institution with over $440 million in assets, serving over 55,000 current and former Alaskan residents. Like similarly situated banks, most of its assets consist of loans and most of its liabilities are deposits. *Interest rate risk* results primarily from potential balance sheet impacts that arise from the timing of interest rate changes relative to the timing of the various loans in its portfolio. Its regulator, the National Credit Union Administration (NCUA), requires that Denali Alaskan measure its interest rate risk by analyzing various interest rate shocks to its balance sheet.

Traditionally, Denali Alaskan used deterministic models to measure this. A major shortcoming of these deterministic models is that they are designed to examine what are believed to be worst-case scenarios and do not provide any information on the probability of these scenarios or yield more likely scenarios for management purposes. So, in 2008, Denali Alaskan began developing a stochastic interest rate generator to forecast both interest rate changes and their impacts on the balance sheet.

The stochastic models that Denali Alaskan developed all use Monte Carlo simulation based on a combination of historical data and management and expert judgment. Fifty years of interest rate data (with information on both the size and speed of changes) have been incorporated, and several subjective judgments are entered concerning factors such as new business development. The models have been back-tested to the significant interest rate shock in 1994–1995, as well as for more recent changes to

* We thank Robert Teachworth, Eric Bingham, and Dale Fosselman for their input to this case.

the balance sheet. Both the CEO and CFO say that the stochastic models provide an increased "comfort and insight level" in their decision making. They feel that forecast ranges are better to communicate than single point estimates.

A final testament to this modeling approach is that Denali's stochastic interest rate generator has become a new business itself; it is considering leasing it to other credit unions, just as American Airlines developed the SABRE reservation system for internal use and then eventually spun it off in 2000 (when it was more valuable than the airline that spawned it).*

2.7 Discrete Event Simulation

This completes our initial foray into Monte Carlo simulation. We have seen how to set up and interpret a simulation, including the use of Tornado Charts and SimTables. The next chapter will show how to use Objects to facilitate building simulation models and to provide additional simulation capabilities.

We end this chapter with a note about types of simulation models that may not be amenable to Monte Carlo simulation within spreadsheets. When you are interested in modeling a project where it is important to simulate the chronological sequence of events (for example, ships' movements in a harbor or patients' stays in a hospital), typically this can best be modeled using discrete event simulation software. Such software packages generally have a "clock," an event list, and a random-number generator. The general idea of discrete event simulation is that a clock starts and, after a random amount of time, events will happen that will have particular effects. For example, a simple problem would be a queue at a bank where customers randomly arrive, and it takes a random amount of time to help each customer. Similarly, patients with different diseases and severity and random needs for treatment arrive randomly in a hospital, but the number of beds is constrained.

While some simple *discrete event* models can be modeled within Excel (possibly with some VBA), modeling such problems in Excel is typically more difficult since it does not have a time tracker or an easy way of keeping track of all the possible events. While for fairly simple discrete event problems some Excel logic (IF statements and so on) can be used, once there are several different events possible, implementing this in Excel quickly becomes too cumbersome. In such cases, dedicated software packages, such as Simul8®, Anylogic®, or Arena®, are much better and efficient choices.

* For more details on the history of SABRE, see Hopper, M. D. 1990. Rattling SABRE—New ways to compete on information. *Harvard Business Review* May–June: 118–125.

CHAPTER 2
Exercises

These exercises are a continuation of those from Chapter 1 (in the same order). Starting with your base case model, you are to incorporate the uncertainty described in each exercise and conduct an appropriate Monte Carlo simulation.

2.1 HEALTH INSURANCE CHOICE SIMULATION

Use your health care insurance plan choice model from Exercise 1.1. Health care service utilization varies across individuals for myriad reasons. While the average health care costs (subject to the copayment percentage) are \$7,000 per year, assume that the actual costs per person have a mean of \$7,000, but follow a lognormal distribution with a standard deviation of 10,000. Average office visits are three per person per year, but these follow a Poisson distribution (with an average of three visits per year). In the United States, health care premiums are tax exempt. (That is, these premiums are not subject to income taxes; assume a marginal tax rate of 35%.) Simulate the cost of the three health care plan options from Exercise 1.1, recognizing the tax benefits and without including the tax benefits. Estimate the probability of a 35-year-old subscribing to each plan, both with and without tax benefits. (This can also be interpreted as the proportion of 35-year-olds that will subscribe to each plan.) Write a paragraph describing your findings.

2.2 ELECTRIC VEHICLE COST SIMULATION

Use your hybrid vehicle model from Exercise 1.2. Gasoline and electricity prices are highly uncertain. Investigate how this uncertainty impacts the break-even additional price for the hybrid engine. Assume that annual change in gasoline price follows a normal distribution (mean = 2%, standard deviation = 25%). Annual changes in electricity prices follow a normal distribution (mean = 6%, standard deviation = 10%). What is the mean break-even price for the hybrid engine? What is the probability that the hybrid engine is more expensive in terms of net present value?

2.3 EDUCATIONAL INVESTMENT SIMULATION

Use your model for estimating the NPV from Exercise 1.3. One source of uncertainty is the possibility of becoming unemployed. High school graduates have an unemployment rate of 9.2%, while those with college degrees (or higher) average 4.8% (use a Bernoulli distribution to model whether or not a person is employed). Assume that, each year, these represent the probability of being unemployed and that a spell of unemployment lasts exactly 1 year. Also assume that each year's chance of being unemployed is independent of all other years and ignore unemployment compensation in this exercise.

As further sources of uncertainty, consider the starting salary to have a normal distribution with the means given in Exercise 1.3 and standard deviations equal to 20% of the means. Simulate NPV for the two investment decisions at two discount rates, 6% and 12%. (These rates are indicative of the difference between student federal loans and private student loans.) What is the probability that the NPV is positive for each investment (at each of these discount rates)?

2.4 CUSTOMER LIFETIME VALUE SIMULATION

Use your LTV model from Exercise 1.4. One source of uncertainty—the retention rate—was already presented in Exercise 1.4. Rather than assuming that the average retention rate is experienced with certainty, simulate the actual retention of customers (use the Binomial distribution for this). Also, assume that the actual retention rate is not known precisely, but follows a PERT distribution with uncertainty of ±5% around each of these rates. Provide 90% confidence intervals for the LTV after 1, 2, and 3 years.

2.5 NETWORK ECONOMICS SIMULATION

Consider your network economics model from Exercise 1.5. There are two major sources of uncertainty that should be accounted for. For a new communications technology, the costs will initially not be known with certainty. Assume that the costs may range between $3 and $8 (with a most likely value of $5, following a PERT distribution). Also, the intrinsic valuations can only be estimated from market research. Assume that the intrinsic valuation function (100(1 – F)) provides an expected estimate, but that the real valuations must be multiplied by an uncertainty factor. This factor will follow a normal distribution with a mean of 1 and a standard deviation of 0.2. Determine 90% confidence intervals for the critical mass and saturation levels.

2.6 PEAK LOAD PRICING SIMULATION

In Exercise 1.6, we saw that peak load pricing can increase profitability by leveling use between peak and off-peak periods (thereby reducing the need for expensive capacity). In this exercise, we examine how uncertain demand influences our results. Suppose that the peak and off-peak demand levels from Exercise 1.6 represent the average hourly demands during these times. Represent the uncertainty by multiplying the average demand by a normal distribution (with mean = 1, standard deviation = 0.25). Each hourly demand will be random and assume that each varies independently of the other hours. Compare the average daily profitability for the single price (50) scheme and the peak-load pricing (75, 25) scheme. What is the probability that peak-load pricing will be more profitable than the single price scheme? Produce a graph that shows how hourly demand levels fluctuate during the day under the two pricing schemes.

2.7 PROJECT MANAGEMENT SIMULATION

Use the project management model from Exercise 1.7. Projects are notorious for running over budget and taking longer than anticipated. Suppose that each activity time from Table 1.3 is uncertain and follows a PERT distribution with the table numbers being the most likely values, but with minimum values = 80% of the table entries and maximum values = 200% of the table entries. (Typically, more things cause a task to take more time rather than less.) Calculate the expected time to project completion and provide a 90% confidence interval for the time to completion. What is the probability that the time will exceed 13 weeks (i.e., the answer from Exercise 1.7)?

2.8 HOLLYWOOD FINANCE SIMULATION

Use your model for the profitability of an average Hollywood movie from Exercise 1.8. The variability of movie profitability has received much study and is the subject of plenty of casual conversation.* Suppose that the costs (production plus marketing) are uncertain and follow a PERT distribution (minimum = $50 million, mode = $100 million, maximum = $450 million). Movie production can take anywhere from 6 to 24 months (following a uniform distribution). Domestic box office receipts also follow a PERT distribution (minimum = $0, mode = $0, maximum = $200 million). Note that this assumption reflects the fact that many movies never run in theatres at all. (We will ignore the fact that movies with higher costs will generally have higher box office receipts. Correlation of uncertainties such as this is the subject of Chapter 7.) Home video sales are still based on domestic box office receipts, but follow a PERT distribution (minimum = 20%, mode = 80%, maximum = 180%). Hit movies (defined as those with greater than $100 million in domestic box office receipts) earn $25 million from premium TV and $25 million from free TV—higher than the figures used in Exercise 1.8.

Estimate the internal rate of return for films. Use the Excel function XIRR (for monthly rates of return) or IRR (for annual returns), but note that IRR or XIRR will produce an error if the undiscounted positive cash flows are sufficiently less than the investment costs.† In your model, identify how often the IRR does not exist. Using the percentile distribution of the IRR (when it does exist), determine the approximate fraction of films that earn a negative return. What fraction earns returns above 30%? (You can compare these fractions to those identified in the footnote. They will not be the same, but they should be qualitatively similar.)

* For some data, see Ferrari, M. J. and A. Rudd. 2008. Investing in movies. *Journal of Asset Management* 19:1. They estimate that 62% of films earn a negative return and 26.6% earn over a 30% return on investment; the average is 3%.

† The IRR (XIRR) is the discount rate that would make NPV = 0, and no positive discount rate will do this under these circumstances. XIRR can estimate a monthly rate of return, but note that the XIRR function requires a date column as an input, as well as a series of values. The IRR and XIRR functions do produce modest negative values, but returns an error if the solution value is too negative. You can use the function ISNUMBER (or ISERROR) to indicate whether or not the IRR formula has returned a number or an error. Also, the *Stats* button in the ModelRisk output window reports the number of errors from the simulation.

3

Modeling with Objects

LEARNING OBJECTIVES

- Learn how to use ModelRisk Objects.
- Use Objects to simplify model construction, troubleshooting, and analysis.
- Use Objects to obtain results without having to run simulations.
- Learn how to model decision problems involving aggregating uncertain frequency of events with uncertain severity of consequences.
- Learn how to model frequency/severity problems correctly without the use of Objects.
- Use Objects and Aggregate distributions to estimate total losses for a portfolio of insurance policies (or similarly structured problems).
- See how the Deduct distribution can be used to model insurance deductibles and/or maxima (or analogous features in similar models).
- See how to model multiple (possibly related) frequency/severity distributions together.
- See how parameter uncertainty (lack of knowledge about parameters in our models) increases our total uncertainty.

3.1 Introductory Case: An Insurance Problem

Insurance companies exist because of the fact that people and companies are generally averse to risk. The essence of the business is that insurance is offered to a population of people that are potentially exposed to the risk of events that could have adverse financial consequences. By pooling the risks of a large number of people who are risk averse, the insurance company can offer an insurance policy at a price to the client that will cover its expected losses and administrative costs, and clients will be willing to pay this premium to avoid the financial consequences of their individual potential losses. Of course, things get complicated when we consider that individuals vary in their riskiness and their attitudes to risk and that the offer of voluntary insurance may itself impact the risks people take (known as moral hazard), as well as affect the likely purchasers of an insurance policy (i.e., adverse selection). These are all interesting complications, but for an insurance company to be able to offer and price insurance policies, it needs to be able to estimate both the *frequency* of losses and the *severity* of these losses when they occur.

Our initial example is based on California auto insurance data for the 1999–2003 time period. We will ignore details such as different risk categories and focus on one type of damage: collision damage. According to the California Department of Insurance, the probability of a collision claim was .07703 per collision damage insurance policy per year, and the average damage claim was $3,321 with a standard deviation of $2,742.* We further assume that the administrative costs (including profits) are 10% of the expected losses of each policy. Then, assuming that there is no further data on how risks vary across individuals (e.g., due to age, driving records, etc.) and that each individual's risk of a claim is independent, what annual collision premium must be charged for this insurer to break even?

3.2 Frequency and Severity

The circumstances in this problem are not unique to the insurance industry. The essential features are that there is an unknown frequency of something occurring and that, every time it occurs, the impact has a certain (unknown) impact. There are numerous examples of situations where both frequency and severity distributions play an important role in understanding risk.

Film studios generally have a number of films in production, only some of which will be released to theatres (frequency). The released films will always vary in their box office receipts (severity). Publishers, music labels, pharmaceutical research, and drilling oil wells undergo similar experiences. Extended warranties offer consumers protection against the possible need for repair as well as the varying size of these repairs. A gambling casino faces uncertainty in the number of winners as well as the variability of the size of the wins. The exercises at the end of this chapter further illustrate the diverse set of situations that share the common features of an unknown frequency and severity.

* The 2008 California Private Passenger Auto Frequency and Severity Bands Manual, Policy Research Division, California Department of Insurance, May 2008. The claim frequency is stated on the basis of exposure years, so this is not the same as the frequency of a claim per driver since some drivers will have more than one claim. We will examine this later in the chapter, but for now we assume that each driver has zero or one claim per year, so .07703 can be considered the probability that an individual driver will have a collision claim during a year.

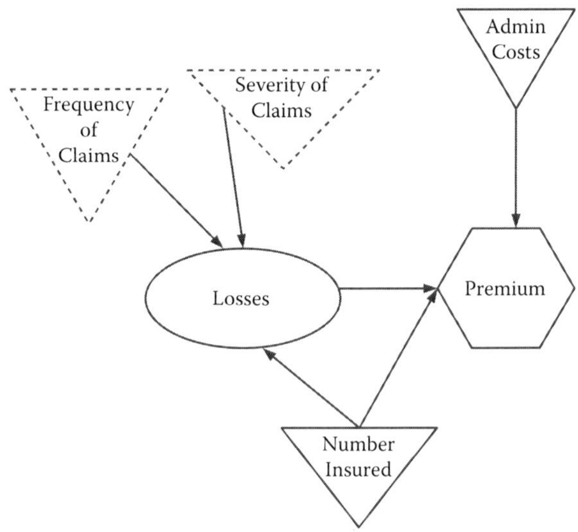

FIGURE 3.1
Insurance visualization.

Both frequency and severity may be modeled using different probability distributions. But care must be taken when these are combined together to estimate the total costs or a mistaken risk assessment will result. We will illustrate this by returning to the insurance problem, which is conceptually shown in Figure 3.1.

To estimate the losses that the insurance company will need to pay out (either total losses or per policy holder), we need to estimate how many claims there will be and, when claims occur, how large each will be. For now, we assume that each insured person will experience either no collision claim or exactly one claim with a probability of .07703. This is modeled for each person as a Bernoulli distribution, shown in Figure 3.2.

Bernoulli distributions apply whenever there are two possible outcomes, a zero or a one, and a probability is attached to each of those outcomes. (The probability for a one is commonly referred to as the "probability of success" or "p.") For our first model, we will assume that there are only 100 insured drivers (i.e., policies) and that the severity of claims follows a Lognormal distribution, as shown in Figure 3.3.

Lognormal distributions are one of several often used to model insurance losses. (Note how the distribution is positively skewed.) Most losses are relatively small, but occasionally a large loss occurs, which generates a long right tail of the severity distribution. Figure 3.4 shows our initial model. (We have hidden the rows for drivers 4 through 98 and placed the final calculations at the top of columns E and F for ease of display.) This is a straightforward model where each driver is assigned a Bernoulli distribution to determine if there is a claim and a Lognormal distribution to determine the severity

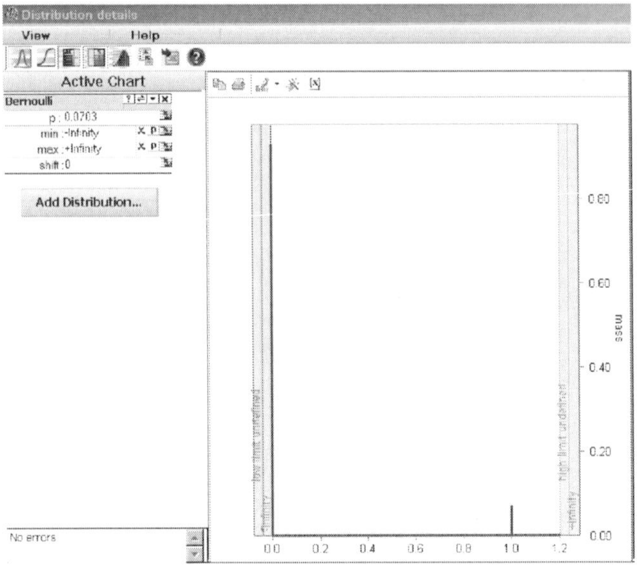

FIGURE 3.2
The Bernoulli distribution window.

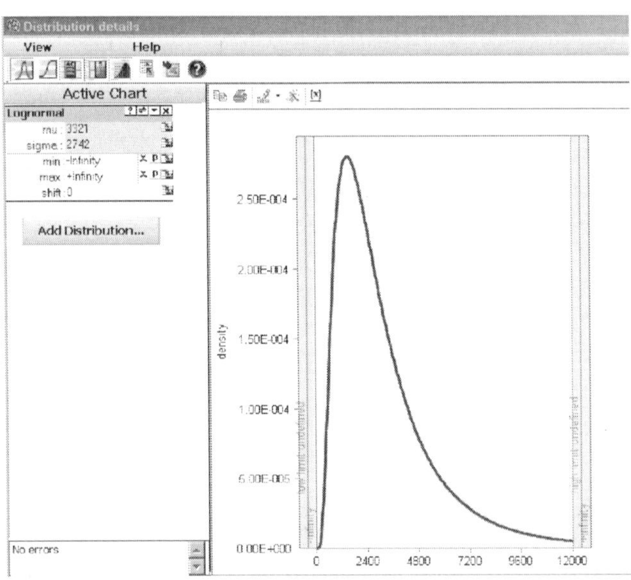

FIGURE 3.3
The Lognormal distribution window.

	A	B	C	D	E	F
1	Parameters				Total Claims	=SUM(D10:D109)
2	Number insured	100			Admin Costs	=B6*F1
3	Claim frequency	0.07703	Bernoulli Distribution		Total Cost	=F1+F2
4	Severity mean	3321	LogNormal Distribution		Premium	=VoseOutput("Premium1","$")+F3/B2
5	Severity std dev	2742			99.9th percentile	=VoseSimPercentile(F4,0.999)
6	Administrative Costs	0.1				
7						
8	Calculations					
9	Driver/Insured	Claim?	possilbe amount	actual claim		
10	1	=VoseBernoulli(0.07703)	=VoseLognormal(3321,2742)	=B10*C10		
11	2	=VoseBernoulli(0.07703)	=VoseLognormal(3321,2742)	=B11*C11		
12	3	=VoseBernoulli(0.07703)	=VoseLognormal(3321,2742)	=B12*C12		
108	99	=VoseBernoulli(0.07703)	=VoseLognormal(3321,2742)	=B108*C108		
109	100	=VoseBernoulli(0.07703)	=VoseLognormal(3321,2742)	=B109*C109		

FIGURE 3.4
The basic insurance model.

when a claim occurs. Note that we model the possible amount of a claim whether a driver has a claim or not, but the actual total claims only count claim amounts for drivers where column B shows that a claim has occurred (i.e., a "1").

The example model is using the Excel option to "Show Formulas."* If you display the formula results rather than the formulas and press the F9 (recalculate) key, you will see that most drivers do not file claims in any individual trial of a simulation. In fact, around ±7 of the 100 drivers typically will file a claim. Running 10,000 simulations (with a random seed of zero) provides results for the insurance premium shown in Figure 3.5.†

Again, notice the "fat" right tail of potential losses, indicating that there is potential for very high losses but that the probability is relatively low. The total expected losses plus administrative costs and profits from selling the 100 insurance policies is $280.09 per policy‡ (found on the Stats display of the ModelRisk results window, which can be added to the histogram using the Display Options dialog) and the 80% confidence interval shown in Figure 3.5 goes from $130.49 to $450.81. Typically, an insurance firm would be most interested in understanding the upper tail of the total loss distribution to better understand how much capital they would have to reserve in order to be (for example) 99.9% sure that the amount reserved is sufficient to cover the losses in a particular year. Thus, we could also examine some of the upper percentiles. (Percentile 99.9 is $826.08 per policy.)

Of course, insurance companies typically have many more than just 100 policies, and therefore we would like to examine a larger model. We would also like to relax some of our assumptions, but it would be nice to find a

* This can be accomplished by using Ctrl + ~ or by using the command in the Formula Auditing box under the Formulas tab in Excel 2007 or later versions.

† It is more common to look at the total loss distribution, and this is what an insurance company would do. In this chapter, we express things on a per-policy basis because total losses depend on the scale of the model (the number of insured drivers), which we have arbitrarily assumed, and because it is an easier quantity to which individual drivers can relate.

‡ You may have noticed that this is indeed very close to .07703 * $3,321 + 10% = $281.40. If we had run more trials, our Monte Carlo approximation would have gotten closer and closer to this amount.

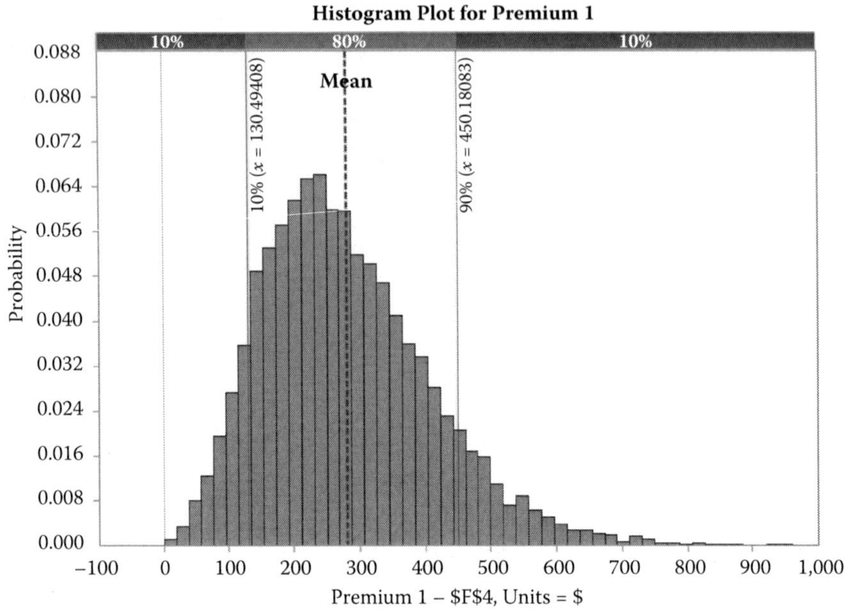

FIGURE 3.5
Initial insurance premium results.

simpler way to model this problem first. Just imagine trying to change the number of insured drivers to 50,000: It is not complicated to copy the rows further down (and adjust the summation formula), but it is unwieldy, unnecessary, and makes for a much slower model to simulate.

A little probability theory helps. If we have a number (n) of independent Bernoulli trials, the number of successes (out of n) will follow a Binomial distribution. The Binomial distribution is based on the number of trials and the (constant) probability of success per trial. Figure 3.6 shows the Binomial distribution for our 100 policy model. (We will increase the number of policies shortly.)

As Figure 3.6 shows, the highest probability is to get seven or eight claims out of 100 people (recall that the probability of a claim is .07703), but six and nine are also fairly common outcomes. Notice that we virtually always have at least one claim and almost never will get more than 16 claims from 100 people. By using the Binomial distribution, we can simplify our model, as shown in Figure 3.7.

This is a very simple spreadsheet, indeed! We only need one cell to simulate the number of claims out of 100 people, so we replace 100 rows of our spreadsheet with a single row. But, be careful because there is a problem. We will mark Premium2 as a second ModelRisk output and run the simulation again. Overlaying our two simulation output cells produces the chart shown in Figure 3.8.

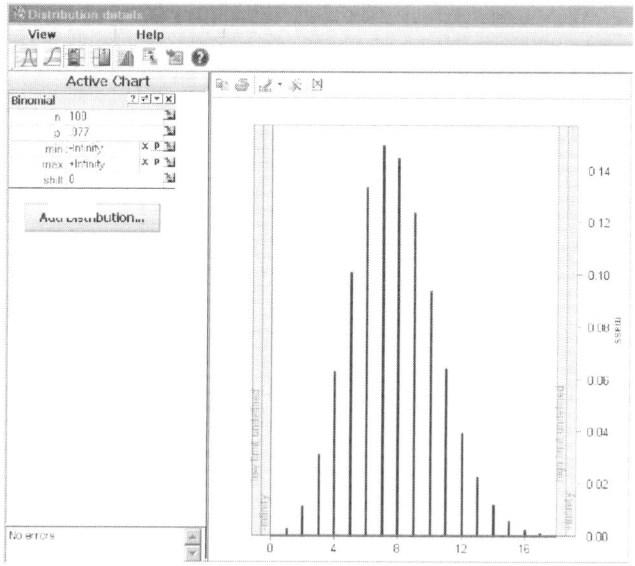

FIGURE 3.6
The Binomial distribution window.

We have chosen the box plot display, which shows the middle 50% of the simulations (the box) and the first and fourth quartiles (called "the whiskers") with a number of outliers. It is easy to see that the distributions look quite different; our simpler model (Premium2) produces much more variable (wider) results. From the Stats display in the results window, we find that the means are close ($280.10 versus $278.12), but the standard deviation for our new simple model ($261.16) is almost twice that in our original model ($128.73). What has gone wrong?

The problem is that our second and smaller model is correctly modeling the frequency—the number of claims out of 100 people—but it is not correctly capturing the severity distribution (i.e., the variability in the size of claims). By using a single cell for the Claim Amount (the cell with the Lognormal distribution), our model is assuming that every claim we simulate with the frequency distribution is for exactly this amount of the severity distribution. The claim amount is different during each trial of the simulation; however, that amount is applied to each claim of that particular trial. While the correct Lognormal distribution is being simulated, it is only being simulated once and then applied to all of the claims that are made.

This is simulating an unrealistic situation where every claim for the 100 drivers during the year has identical severity. The probability of this happening is vanishingly small. The original model correctly permitted each claim to vary independently of the others. This results in much less uncertainty (i.e., a narrower distribution) in the resulting total costs, since some

	E	F	G	H	I	J	K
1	Total Claims	=SUM(D10:D109)					
2	Admin Costs	=B6*F1					
3	Total Cost	=F1+F2					
4	Premium	=VoseOutput("Premium 1","$")+F3/B2					
5	99.9th percentile	=VoseSimPercentile(F4,0.999)					
6							
7							
8		Alternative Calculation					
9		Number of Claims	Claim Amount	Total Claims	Admin Costs	Total Cost	Premium 2
10		=VoseBinomial(B2,B3)	=VoseLognormal(B4,B5)	=F10*G10	=B6*H10	=H10+I10	=VoseOutput("Premium 2","$")+J10/B2

FIGURE 3.7
A one-row insurance model.

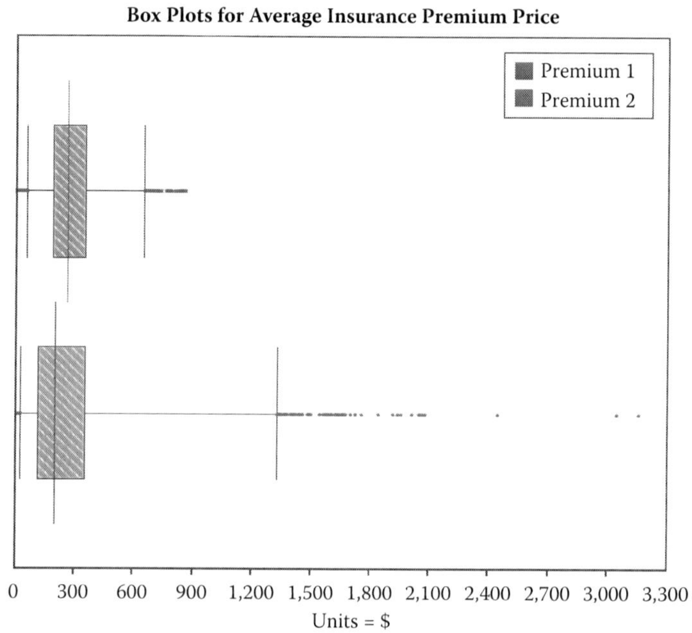

FIGURE 3.8
Results from the (incorrect) one-row insurance model.

large claim amounts will be offset by others with small claim amounts. The simple model incorrectly does not permit large and small claims to offset each other; all claims are forced to be the same size in each simulation trial.

Fortunately, there is a way to fix this problem and, at the same time, keep the model simple through the use of *Distribution Objects*.

3.3 Objects

Among Monte Carlo simulation tools, Objects are unique to ModelRisk. They serve several purposes; they can simplify the construction of models as well as provide several additional capabilities. One of these capabilities will permit us to produce a simple and correct model for the insurance problem. The next chapter will show additional capabilities that Objects provide.

When you define a distribution in ModelRisk, such as the Binomial distribution used to model the number of claims, you have a choice of how to place the distribution into your spreadsheet. Thus far, we have pasted the distribution directly into the spreadsheet so that it generates a random sample from the distribution each time the model is recalculated (as when the F9 key is pressed). But there is also an option to paste the distribution as an

FIGURE 3.9
Paste to Excel button.

Object. Figure 3.9 shows the Paste to Excel button on the Distribution dialog window, under which these paste options appear.

The ModelRisk functions VoseBinomial(100, .07703) and VoseBinomial-Object (100,.07703) both represent identical distributions.* They represent the same Binomial distribution (100 trials, each with a probability of .07703 per trial). The VoseBinomial(100, .07703) puts the simulated value into the spreadsheet, while the Object does not. The Object represents the full distribution but will not produce a numerical value until you ask it to provide something, such as a random value or a parameter (such as a percentile). For example, if VoseBinomialObject(100, .07703) is entered into a cell (e.g., A1) and then the function VoseSimulate(A1) is entered into another cell (B1), then exactly the same result is obtained in cell B1 as if VoseBinomial(100,.07703) had been used in that cell—a random simulated value from this distribution. It may seem like an extra step since we can get the same result by putting the distribution into a cell as when using the Distribution Object and then having to simulate it in a separate cell. The use of Objects can, however, greatly simplify a model as well as provide it with more capability.

3.4 Using Objects in the Insurance Model

To see how it simplifies model construction, we have used Objects to build the insurance model for the 100 separate insurance policies, as shown in Figure 3.10. This model will produce the same results as our original (correct, but large) model. Notice that the Calculation section no longer has any distributions in it. The two distributions are entered as Objects in column C and the Calculation section simply refers to these Object cells using the VoseSimulate function. (Even though the function looks the same on each row, each use of the VoseSimulate function gives a different random simulation from the Distribution Object.) While this model still contains a row for each driver, it has improved our model in two ways. First, all of the assumptions are in the parameter section. Troubleshooting and reviewing

* Note that they will appear differently in the spreadsheet. The second function will look like VoseBinomial(100,.07703), but if you click on the cell, the formula bar will show it to be VoseBinomialObject(B2,B3).

	A	B	C	D
1	Parameters			
2	Number insured	100		
3	Claim frequency	0.07703	Bernoulli Distribution Object	=VoseBernoulliObject(B3)
4	Severity mean	3321	LogNormal Distribution Object	=VoseLognormalObject(B4,B5)
5	Severity std dev	2742		
6	Administrative Costs	0.1		
7				
8	Calculations			
9	Driver/Insured	Claim?	possilbe amount	actual claim
10	1	=VoseSimulate($D$3)	=VoseSimulate($D$4)	=B10*C10
11	2	=VoseSimulate($D$3)	=VoseSimulate($D$4)	=B11*C11
12	3	=VoseSimulate($D$3)	=VoseSimulate($D$4)	=B12*C12
108	99	=VoseSimulate($D$3)	=VoseSimulate($D$4)	=B108*C108
109	100	=VoseSimulate($D$3)	=VoseSimulate($D$4)	=B109*C109

FIGURE 3.10
The basic insurance model using Objects.

and changing the model are more straightforward. Second, it is easier to perform "what if" analyses. Suppose that we want to change severity distribution from a Lognormal to some other distribution. Our original model would require us to adjust the formulas in column C for each driver. Our new model only requires us to change one cell (D4) and the rest of the model stays intact.

A greater benefit, however, is that the use of Objects will also give us the opportunity to model the total losses of all 100 households within a single cell. To accomplish this, we use the Aggregate menu on the ModelRisk ribbon. The Aggregate functions sum (or aggregate) a number of independent and identically distributed ("iid") random variables. For example, if drivers' collision damage distributions are iid, then the total damage is a random variable that is obtained by summing these individual distributions, one for every accident. We could model each distribution separately and then add them together as we did in the original model, or we can use an Aggregate function to do this for us.

There are a number of Aggregate choices; for now, just use the Aggregate Monte Carlo function. When you select this function from the Aggregate drop-down menu, you get a dialog box that asks you to specify a frequency distribution and a severity distribution, as shown in Figure 3.11.

You can choose these distributions by clicking on the button for choosing from the list of available distributions. You will also find, under the Select Distribution button, that there are categories of distributions for claim frequency and claim size. Not all distributions available in ModelRisk can be used in Aggregate modeling, but many can, including the Binomial as the frequency distribution (always discrete) and Lognormal as the severity distribution (always a continuous distribution) from our insurance model.

Use of the Aggregate Monte Carlo capability in ModelRisk requires that the claim frequency and claim size distributions be entered as Objects. This is because ModelRisk will correctly sample from the claim frequency

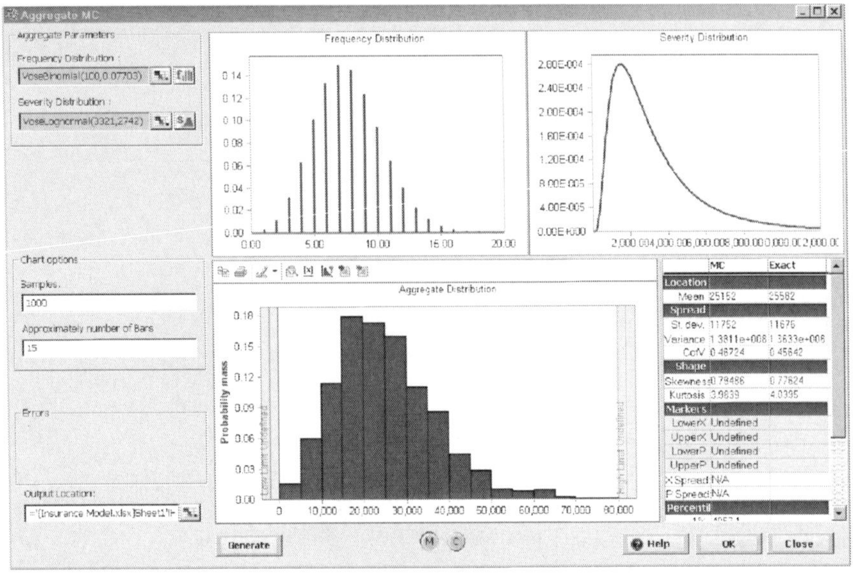

FIGURE 3.11
The Aggregate Monte Carlo distribution window.

distribution and then simulate the variability of the claim sizes. The Distribution Objects can be entered manually into the dialog box, selected via the Select Distribution button, or defined by referencing an Object entered into another cell of the model.

We use the frequency (D18) and severity (E4) Objects in the Aggregate Monte Carlo dialog box as shown in Figure 3.12 and place the output in cell E18 of our spreadsheet. We now have a third version of our insurance model (Insurance3b.xlsx) shown in Figure 3.12. To verify that this new model works correctly, we calculate the required collision insurance premium as Premium3. Running the simulation and overlaying the original correct Premium1 (obtained by modeling each driver separately), the new correctly modeled Premium3 (using Objects and the AggregateMC function), along with the incorrect Premium2 (based on each claim being assigned the same severity), gives Figure 3.13.

Notice that the distributions for Premium1 and Premium3 are quite close* (almost identical percentiles), but Premium2 has a much wider (and incorrect) distribution. Unlike our previous failed attempt to reduce the model to a single row (Premium2), this single row (using Objects and the Aggregate capability) correctly models the insurance problem.

To appreciate the power of Objects fully, we will now model 1,000,000 drivers. It is not an attractive option to model each driver separately, so we will use Objects and model the claims and premium with a single row.

* There are slight differences, which would get smaller if we were to run more iterations.

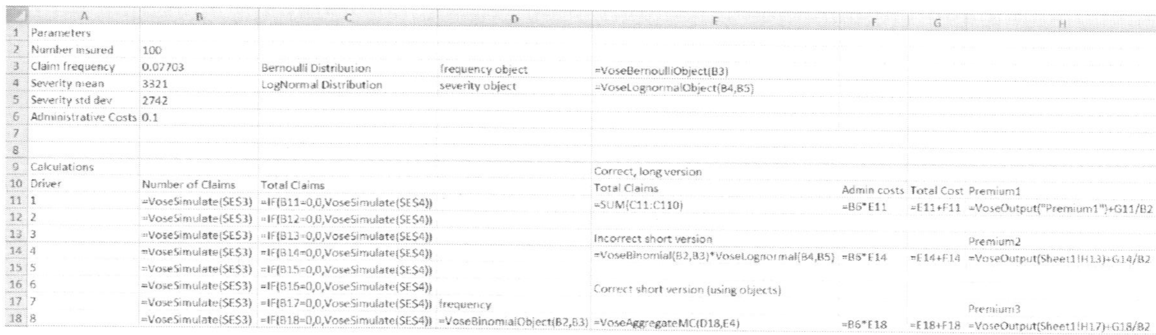

	A	B	C	D	E	F	G	H
1	Parameters							
2	Number insured	100						
3	Claim frequency	0.07703	Bernoulli Distribution	frequency object	=VoseBernoulliObject(B3)			
4	Severity mean	3321	LogNormal Distribution	severity object	=VoseLognormalObject(B4,B5)			
5	Severity std dev	2742						
6	Administrative Costs	0.1						
7								
8								
9	Calculations				Correct, long version			
10	Driver	Number of Claims	Total Claims		Total Claims	Admin costs	Total Cost	Premium1
11	1	=VoseSimulate($E$3)	=IF(B11=0,0,VoseSimulate($E$4))		=SUM(C11:C110)	=B6*E11	=E11+F11	=VoseOutput("Premium1")+G11/B2
12	2	=VoseSimulate($E$3)	=IF(B12=0,0,VoseSimulate($E$4))					
13	3	=VoseSimulate($E$3)	=IF(B13=0,0,VoseSimulate($E$4))		Incorrect short version			Premium2
14	4	=VoseSimulate($E$3)	=IF(B14=0,0,VoseSimulate($E$4))		=VoseBinomial(B2,B3)*VoseLognormal(B4,B5)	=B6*E14	=E14+F14	=VoseOutput(Sheet1!H13)+G14/B2
15	5	=VoseSimulate($E$3)	=IF(B15=0,0,VoseSimulate($E$4))					
16	6	=VoseSimulate($E$3)	=IF(B16=0,0,VoseSimulate($E$4))		Correct short version (using objects)			
17	7	=VoseSimulate($E$3)	=IF(B17=0,0,VoseSimulate($E$4))	frequency				Premium3
18	8	=VoseSimulate($E$3)	=IF(B18=0,0,VoseSimulate($E$4))	=VoseBinomialObject(B2,B3)	=VoseAggregateMC(D18,E4)	=B6*E18	=E18+F18	=VoseOutput(Sheet1!H17)+G18/B2

FIGURE 3.12
A correct one-row insurance model.

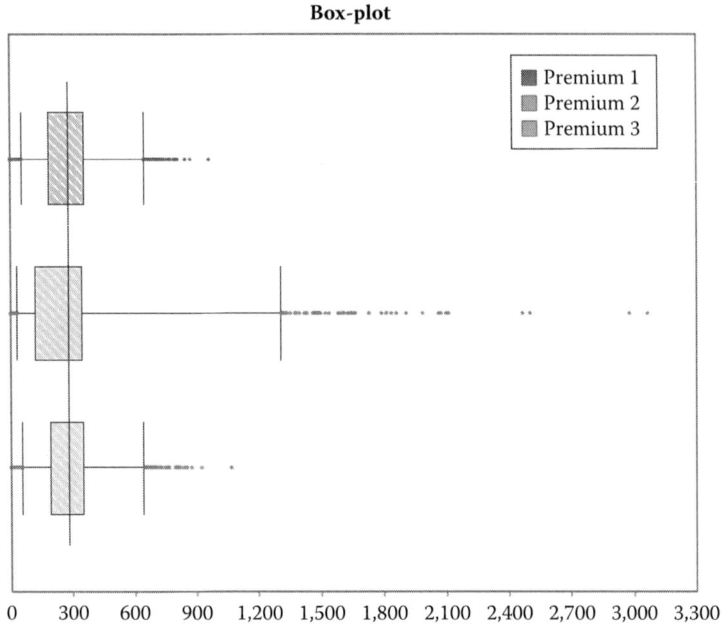

FIGURE 3.13
Results of the basic and one-row insurance models.

Another modification is in order since, with 1,000,000 drivers, some drivers will experience more than one claim, and the Bernoulli distribution only permits outcomes of zero or one. We change the Claim Frequency distribution from a Binomial distribution to a Poisson distribution. In many models structured with a frequency and severity, the frequency distribution is modeled as a Poisson distribution rather than a Binomial one.* In our example, the number of events can be thought of as either a number of trials (100) with a specific probability of occurring—in which case, we would use Binomial(100, .07703)—or as a number of events with an average rate of occurrence per year—in which case, we would use Poisson(100*.07703). Because the rate per year is fairly small, it will be rare to have two accidents per year and therefore the Binomial and Poisson distributions will be very similar and give similar results in this case.

When choosing between the Binomial and Poisson distributions, it is best to consider both the nature of the event being modeled and the form of the

* The Binomial and Poisson distributions are closely related and sometimes the Binomial distribution can be approximated by a Poisson distribution. The Poisson measures the number of successes within a range of exposure (e.g., over a certain time period), while the Binomial measures the number of successes out of a specified number of trials.

available data. If you are modeling an event that has only two possible outcomes and the probability of each outcome can be estimated (like flipping a coin), then the Binomial method is usually the best choice. However, if you are modeling the number of occurrences of an event that happens with a particular rate (for example, the number of customers arriving at a restaurant per day), then the Poisson method is the best choice. The Poisson distribution also allows the number of claims per individual driver to take on integer values other than zero or one. (In this case, since the probability of claim is low—.07703—very few drivers will experience more than one claim in a year.)

To illustrate the convenience of an Object-based model, we will run the model with 100 and 1,000,000 drivers (Insurance3c.xlsx). The results for the two premiums (Premium5, with 100 drivers, and Premium6, with 1,000,000 drivers) are shown in Figure 3.14.

The box plot shows the results for both models, and the histogram shows the results for Premium6 alone. The mean is $281.40 and the 80% confidence interval ranges from $279.69 to $283.10. The box is a great demonstration of how spreading risk across a large population of policies can drastically reduce the risk assumed by the insurance company. The reason the interval is so narrow is that, with 1,000,000 drivers (assuming that they are independent), there are 10,000 times as many drivers and therefore a few more or few less incidents do not have much effect on the total claim size anymore. In fact, more policies tend to cancel each other out, producing a mean loss per insured that spans a much smaller range of values. A much larger risk pool has made the total loss quite predictable.

3.5 Modeling Frequency/Severity without Using Objects

It is also possible to use a small model, without Objects, to estimate total losses accurately by using some additional probability theory. The Central Limit Theorem states that the distribution of a sample mean will approach (as the sample size increases) a Normal distribution with mean equal to the population mean, and standard deviation equal to the population standard deviation divided by the square root of the sample size. Therefore, we can multiply the simulated number of claims by the average claim amount based on the Central Limit Theorem and use the Normal distribution rather than the Lognormal distribution. While this sounds somewhat complicated, it will make a lot more sense when you see it implemented in this example. Figure 3.15 shows this model, again with 1,000,000 drivers, and the result is now named Premium7.

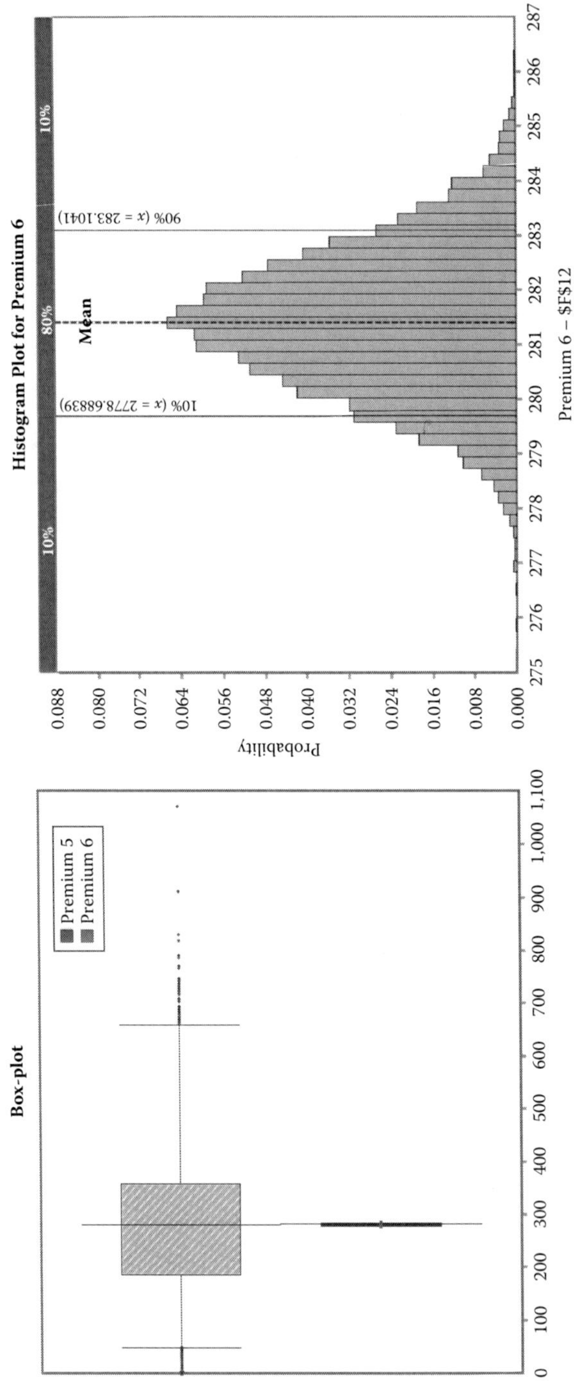

FIGURE 3.14
Insurance results for 1,000,000 drivers.

Parameters						Alternate (correct w/o Objects)
			number insured	100	1000000	1000000
Claim frequency	0.07703	Poisson Distribution	frequency object	=VosePoissonObject($B$3*E2)	=VosePoissonObject($B$3*F2)	=VosePoisson(B3*G2)
Severity mean	3321	LogNormal Distribution	severity object	=VoseLognormalObject($B$4,$B$5)	=VoseLognormalObject($B$4,$B$5)	=VoseNormal(B4,B5/(SQRT(G3)))
Severity std dev	2742					
Administrative Costs	0.1					
			Calculations			
			Total Claims	=VoseAggregateMC(E3,E4)	=VoseAggregateMC(F3,F4)	=G3*G4
			Administrative costs	=$B$6*E9	=$B$6*F9	=$B$6*G9
			Total Costs	=E9+E10	=F9+F10	=G9+G10
			Premium	=VoseOutput("Premium5")+E11/E2	=VoseOutput("Premium6")+F11/F2	=VoseOutput("Premium7")+G11/G2

FIGURE 3.15
Simplified model without using Objects.

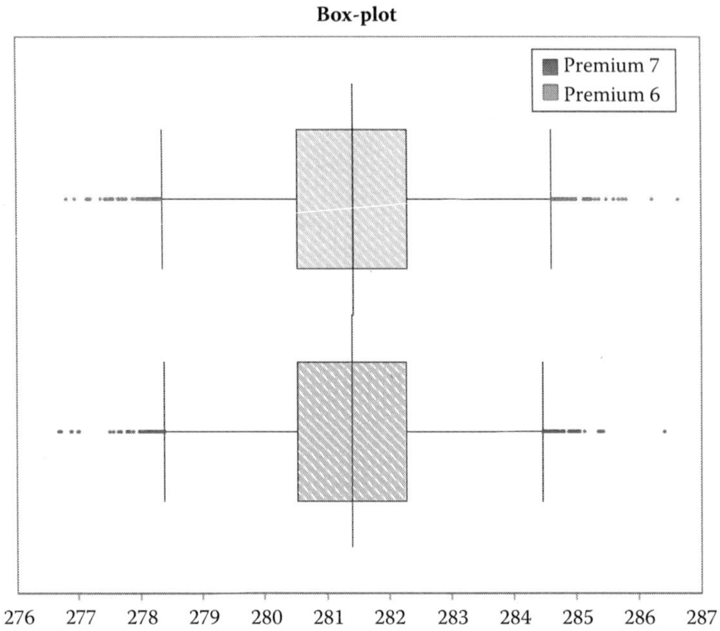

FIGURE 3.16
Results of the corrected simplified model.

Note that the sample size number used in the Normal distribution (cell G4) is not the number of drivers but rather the number of claims. This model gives almost exactly the same results as obtained with Aggregate modeling using the Lognormal Severity function for the individual claim severity, as shown in Figure 3.16.

So, careful application of probability theory can allow frequency/severity models to be built efficiently without needing to model each individual random variable separately and without the use of Objects or Aggregate modeling. However, care must be taken to ensure that the sample size is sufficiently large for this approach. If the underlying severity distribution (the Lognormal, in this case) is fairly skewed, then the sample size (i.e., the number of loss events) should be at least 30 in order for the Normal to be a good approximation of the average severity within the sample. When the frequency probability is low (.07703 in this case), this translates to a substantially larger sample size. (n must be large enough that the probability of getting less than 30 events is quite small; around 525 policies translates to this probability being close to 5%.) Thus, our 100 policy insurance model would not be appropriate for this approach. Of course, with smaller sample sizes (such as 100), it is feasible to construct the spreadsheet to model each policy separately.

3.6 Modeling Deductibles

ModelRisk has additional features useful in insurance modeling (or other models with similar structure). For example, there is a Deduct distribution that permits a claim size distribution to contain a deductible and/or a maximum claim size. (Both features are common in insurance policies.) We could replace the current Lognormal Object for the claim size distribution with the Deduct distribution, as shown in Figure 3.17, for a $500 deductible and no claim maximum.

Our new model (Insurance3d.xlsx) is shown in Figure 3.18. Notice the use of the VoseDeductObject, based on the VoseLognormalObject, but specifying the $500 deductible.

We now run the model (for 1,000 policy holders because it takes a long time to simulate 1,000,000 drivers) and obtain the results shown in Figure 3.19.

The required premium is reduced when the policy has a deductible. The mean premium drops from about $282 to around $239. As one further enhancement to our model, we incorporate a Simtable into our VoseDeductObject model to consider deductibles ranging from $100 to $1,000, as shown in Figure 3.20.

The Simtable is constructed as we described in Chapter 2, using the VoseSimTable function in cell B7 rather than the fixed $500 deductible. Cell C11, which contains the VoseDeductObject, uses cell B12 as the deductible amount. The table is in B13:E23 and will report the means and standard deviations from the simulations (after setting the number of simulations to 10).

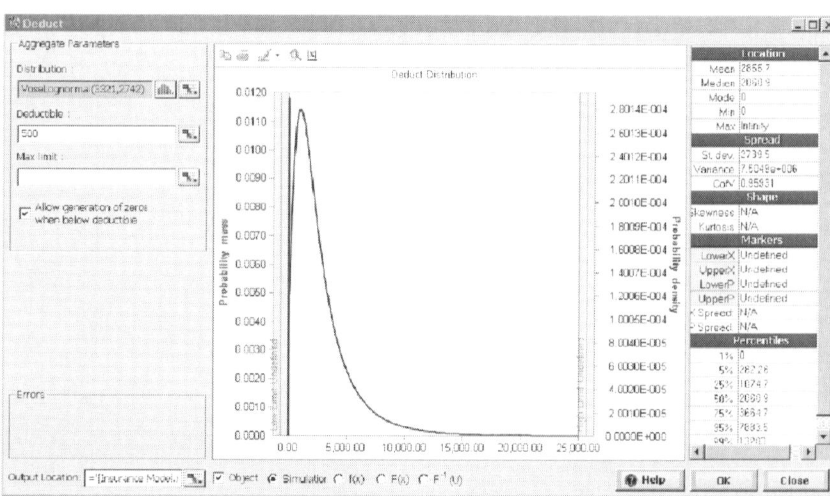

FIGURE 3.17
The Deduct distribution window.

	A	B	C	D	E	F	G
1	Parameters						
2	Number insured	1000					
3	Claim frequency	0.07703	Poisson Distribution				
4	Severity mean	3321	LogNormal Distribution				
5	Severity std dev	2742					
6	Administrative Costs	0.1					
7	Deductible	500					
8							
9	Calculations	Number of Claims	Claim Amount	Total Claims	Admin Costs	Total Costs	Premium
10		=VosePoissonObject($B$2*$B$3)	=VoseLognormalObject($B$4,$B$5)	=VoseAggregateMC(B10,C10)	=$B$6*D10	=D10+E10	=VoseOutput("without deductible","$")+F10/B2
11		=VosePoissonObject($B$2*$B$3)	=VoseDeductObject(C10,B7,,TRUE)	=VoseAggregateMC(B11,C11)	=$B$6*D11	=D11+E11	=VoseOutput("with deductible","$")+F11/B2

FIGURE 3.18
Insurance model using the Deduct distribution.

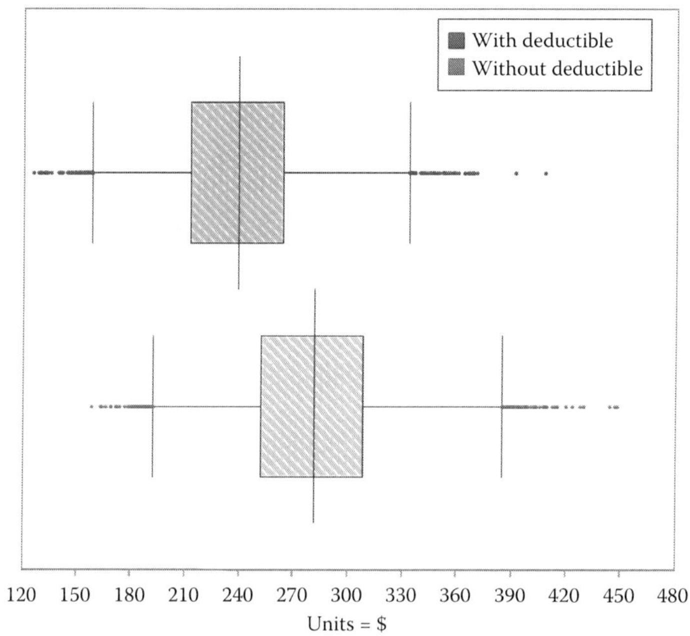

FIGURE 3.19
Comparing results with and without deductibles.

The results for the mean and standard deviation of the premium are graphed in Figure 3.21.

You can see how the mean premium drops as the deductible increases. Not as obvious, perhaps, is that the standard deviation also drops (from $41 to $35 over the range of deductibles). The drop in the standard deviation is mostly the result of the left tail of the distribution being cut off, while the right tail does not change much. It is the right tail that insurers are mostly concerned about, so adding "maximum payouts" per policy is a more powerful way of reducing an insurer's risk.

You may have noticed that there are several types of Aggregate functions available in ModelRisk. We now compare the Aggregate Monte Carlo (AggregateMC) with the Aggregate Fast Fourier Transform (AggregateFFT) methods.* We omit the deductible and forego modeling the drivers individually. The model (Insurance3e.xlsx) is shown in Figure 3.22.

Row 10 contains the Monte Carlo Aggregate simulation, with output labeled Premium8 (in cell F4), and row 11 contains the AggregateFFT method, with output labeled Premium9 (in cell G4). An important distinction between

* Some additional options, which are alternative approximation methods to calculate the total claim size, are available. In addition, there is an option to use a multivariable (AggregateMultiMC) version; this permits modeling of multiple risk classes (which may or may not be correlated).

	A	B	C	D	E	F	G
1	Parameters						
2	Number insured	1000					
3	Claim frequency	0.07703	Poisson Distribution				
4	Severity mean	3321	LogNormal Distribution				
5	Severity std dev	2742					
6	Administrative Costs	0.1					
7	Deductible	=VoseSimTable(C14:C23,,500)					
8							
9	Calculations	Number of Claims	Claim Amount	Total Claims	Admin Costs	Total Costs	Premium
10		=VosePoissonObject($B$2*$B$3)	=VoseLognormalObject($B$4,$B$5)	=VoseAggregateMC(B10,C10)	=$B$6*D10	=D10+E10	=VoseOutput("without deductible","$")+F10/B2
11		=VosePoissonObject($B$2*$B$3)	=VoseDeductObject(C10,B7,,TRUE)	=VoseAggregateMC(B11,C11)	=$B$6*D11	=D11+E11	=VoseOutput("with deductible","$")+F11/B2
12							
13		Simulation	Deductible	Mean Premium	Std dev.		
14		1	100	=VoseSimMean($G$11,B14)	=VoseSimStdev($G$11,B14)		
15		2	200	=VoseSimMean($G$11,B15)	=VoseSimStdev($G$11,B15)		
16		3	300	=VoseSimMean($G$11,B16)	=VoseSimStdev($G$11,B16)		
17		4	400	=VoseSimMean($G$11,B17)	=VoseSimStdev($G$11,B17)		
18		5	500	=VoseSimMean($G$11,B18)	=VoseSimStdev($G$11,B18)		
19		6	600	=VoseSimMean($G$11,B19)	=VoseSimStdev($G$11,B19)		
20		7	700	=VoseSimMean($G$11,B20)	=VoseSimStdev($G$11,B20)		
21		8	800	=VoseSimMean($G$11,B21)	=VoseSimStdev($G$11,B21)		
22		9	900	=VoseSimMean($G$11,B22)	=VoseSimStdev($G$11,B22)		
23		10	1000	=VoseSimMean($G$11,B23)	=VoseSimStdev($G$11,B23)		

FIGURE 3.20
SimTable for deductibles.

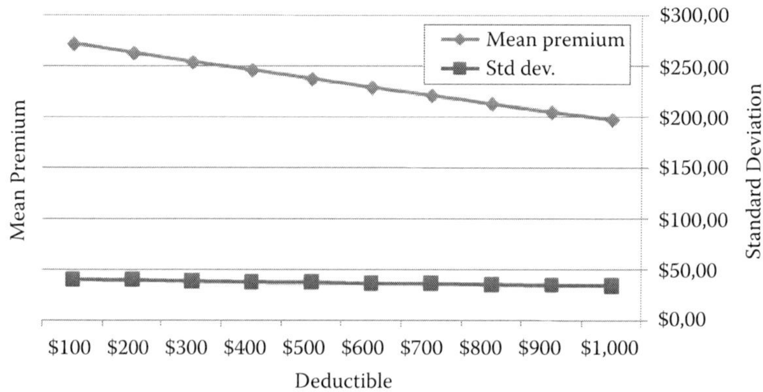

FIGURE 3.21
The effects of deductibles on insurance premiums.

these two Aggregate functions is that the Monte Carlo method actually conducts a true simulation "behind the scenes" to generate the Aggregate distribution, while the FFT method is an approximation that computationally determines the probabilities and the Aggregate distribution. One result of these differences is that the AggregateFFT is available in ModelRisk as an Object, while the AggregateMC is not. There are some desirable advantages to using Objects that make the AggregateFFTObject useful; however, there are also disadvantages in that the FFT method is an approximation, which in some circumstances does not give accurate results (as we shall see).

We run the simulation, first for 100 insured drivers and then for 100,000 (we only need to change cell B2 for this), and get the results shown in Figure 3.23. The results for 100 drivers are quite close, but something is clearly incorrect with the FFT method when looking at 100,000 drivers. As we found earlier, it is expected that the distributions be much narrower with many more drivers; however, the FFT is now generating much different results. To see why the FFT method is inappropriate with 100,000 drivers, look at the dialog box for the AggregateFFTObject. (Highlight cell C11 and click the View Function button* on the ModelRisk ribbon to bring up this dialog.)

As a way to determine if the FFT calculations are providing a good estimate of the Aggregate function, ModelRisk displays both the FFT and exact values for some statistics of the Aggregate function in the lower right of the window, as shown in Figure 3.24.

When we are modeling 100,000 drivers, the values are not similar, which indicates that the FFT method is not a good approximation method to use in this case. Contrast this with the same dialog when the number of

* The View Function button is a convenient way to return to the dialog window for any ModelRisk function.

	A	B	C	D	E	F
1	Parameters			Total Claims	=C10	=C11
2	Number insured	100		Admin Costs	=B6*E1	=B6*F1
3	Probability of a claim	0.07703	Binomial	Total Cost	=E1+E2	=F1+F2
4	Severity mean	3321	LogNormal Distribution	Premium	=VoseOutput("Premium 8","$")+E3/B2	=VoseOutput("Premium 9","$")+F3/B2
5	Severity std dev	2742		99.9th percentile	=VoseSimPercentile(E4,0.999)	=VoseSimPercentile(F4,0.999)
6	Administrative Costs	0.1				
7						
8	Alternative Calculation					
9	Frequency	Claim Amount	Total Claims			
10	=VoseBinomialObject(B2,B3)	=VoseLognormalObject(B4,B5)	=VoseAggregateMC(A10,B10)			
11	=VoseBinomialObject(B2,B3)	=VoseLognormalObject(B4,B5)	=VoseAggregateFFT(A11,B11)			

FIGURE 3.22
The AggregateFFT model.

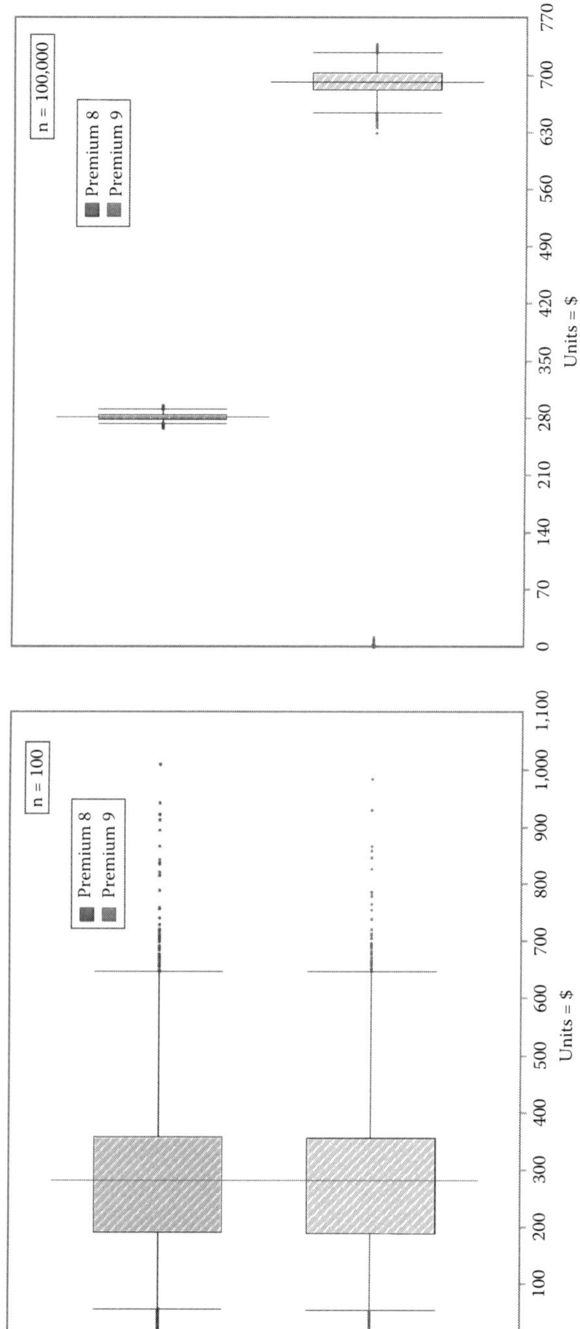

FIGURE 3.23
Correct and incorrect results from the AggregateFFT model.

	FFT	Exact
Location		
Mean	1037.1	2.5582e+007
Spread		
St. dev	29041	3.6923e+005
Variance	8.4337e+008	1.3633e+011
CofV	28.002	0.014433
Shape		
Skewness	39.703	0.024547
Kurtosis	2004.6	3.001

FIGURE 3.24
Verifying the inaccuracy of the AggregateFFT method.

	FFT	Exact
Location		
Mean	25581	25582
Spread		
St. dev.	11675	11676
Variance	1.363e+008	1.3633e+008
CofV	0.45638	0.45642
Shape		
Skewness	0.77449	0.77624
Kurtosis	4.019	4.0395

FIGURE 3.25
Verifying the accuracy of the AggregateFFT method.

drivers is 100 (shown in Figure 3.25). The values are close here, so the FFT approximation can be used.*

3.7 Using Objects without Simulation

Objects, as we have seen, are very useful when constructing models and when implementing Aggregate distributions. In fact, the use of Objects can sometimes even enable us to bypass the need to Monte Carlo simulate at all! We will look at how this can be done using VoseAggregateFFT, but this also applies to any Distribution Object in ModelRisk. For some models, ModelRisk can use the frequency and severity Distribution Objects to calculate different statistics of the total claim distribution directly, without simulation.

Suppose that we want to know the probability that the average costs of each policy to the insurer will be greater than $297, when there are 100 insured drivers. With 100 policies, a premium of $297 corresponds to total claims of $27,000 (and $2,700 of administrative expense). So, the probability that the

* It is not the case that the FFT method always fails for large values of n. There is no general rule regarding when FFT can be used, but you can (and should) always check the dialog box to make sure that the FFT and exact values are close before using the FFT method.

average costs will exceed $297 should be the same as the probability that the total claims will exceed $27,000.

Figure 3.26 shows the last modification to our insurance model (Insurance3f.xlsx). Here we removed the AggregateMC and are modeling only the AggregateFFTObject in cell C10. D10 contains the VoseSimulate function to draw random samples from the AggregateFFTObject and is designated as an output named "Total Claims." D13 contains the ModelRisk VoseSimProbability function, which will display the probability of cell D10 being greater than $27,000 after a simulation has been run.

So, we could run a simulation to obtain this estimate, but it is not necessary if we use the FFT method. We can calculate this probability directly by entering the formula, =1-VoseProb(27000,C10,TRUE), into our spreadsheet. The function VoseProb can calculate the probability directly from the VoseAggregateFFTObject (which is cell C10).* TRUE indicates that we want the cumulative probability, which we subtract from 1 to get our estimate (since the cumulative probability is the probability of getting total claims *less than* or equal to 27,000). Even before running a simulation, this result displays 40.35% in cell D14. If we run the model in Figure 3.26, the simulated result is 40.18%. The FFT result is quite close to the simulation result (both of which are approximations), without the need to simulate.

3.8 Multiple Severity/Frequency Distributions†

Our final insurance model will involve analysis of multiple insurance claim frequency and severity distributions at the same time. The California Department of Insurance Data contains information on a number of auto insurance claim categories: bodily injury, property damage, medical payments, uninsured motorist, collision, and comprehensive. We will use the frequency and severity data for all of these claim types, along with typical deductible and maximum limits for what is called a "standard" policy. We can again take advantage of the VoseDeductObject to include these features. We also use 26% for the non-claims-related expenses and an average premium of $800.‡ Note that premiums vary significantly across drivers and automobiles, but our model does not consider these different levels of risk at all. Figure 3.27 shows our model (Insurance3g.xlsx).

* Note that all Vose functions are fully documented through the Excel function help, as well as through ModelRisk help.

† This section is more advanced and may be skipped without upsetting the continuity of the book.

‡ Both estimates come from the Insurance Information Institute website (www.iii.org), accessed on July 28, 2010.

	A	B	C	D
1	Parameters			
2	Number insured	100		
3	Probability of a claim	0.07703	Binomial	
4	Severity mean	3321	LogNormal Distribution	
5	Severity std dev	2742		
6	Administrative Costs	0.1		
7				
8	Alternative Calculation			
9	Frequency	Claim Amount	Total Claims - Aggregate Object	Total Claims - Simulate
10	=VoseBinomialObject(B2,B3)	=VoseLognormalObject(B4,B5)	=VoseAggregateFFTObject(A10,B10)	=VoseOutput("Total Claims","$")+VoseSimulate(C10)
11				
12			Probability Total Claims > $30,000	
13			From simulation	=1-VoseSimProbability(D10,27000,TRUE)
14			From direct calculation	=1-VoseProb(27000,C10,TRUE)

FIGURE 3.26
Results showing the probability of a premium exceeding $300.

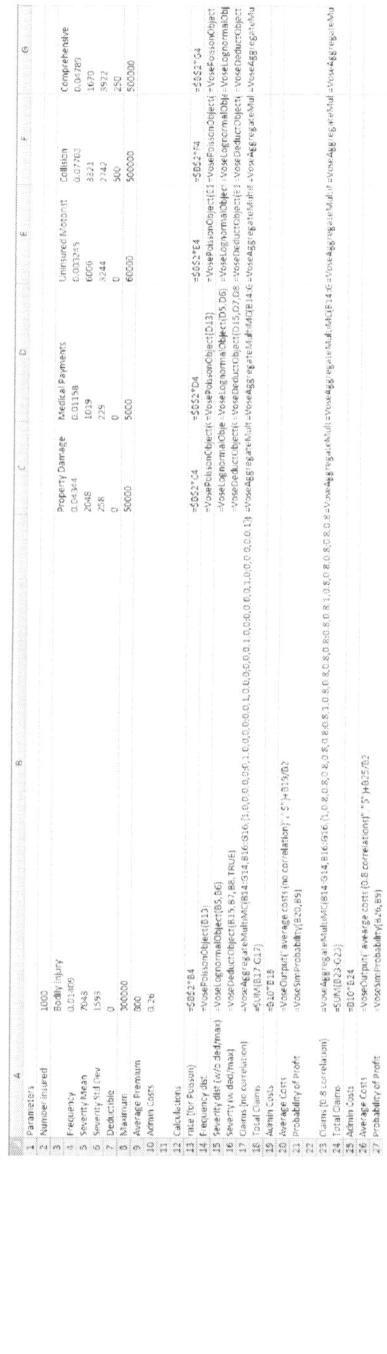

	A	B	C	D	E	F	G
1	Parameters						
2	Number insured	1000					
3		Bodily Injury	Property Damage	Medical Payments	Uninsured Motorist	Collision	Comprehensive
4	Frequency	0.01409	0.04344	0.01158	0.003245	0.07703	0.04789
5	Severity Mean	7043	2048	1019	6006	3321	1670
6	Severity Std Dev	1593	258	229	3244	2742	3922
7	Deductible	0	0	0	0	500	250
8	Maximum	300000	50000	5000	60000	500000	500000
9	Average Premium	800					
10	Admin Costs	0.26					
11							
12	Calculations						
13	rate (for Poisson)	=$B$2*B4	=$B$2*C4	=$B$2*D4	=$B$2*E4	=$B$2*F4	=$B$2*G4
14	Frequency dist	=VosePoissonObject(B13)	=VosePoissonObject(	=VosePoissonObject(D13)	=VosePoissonObject(E1	=VosePoissonObject(	=VosePoissonObject
15	Severity dist (w/o ded/max)	=VoseLognormalObject(B5,B6)	=VoseLognormalObje	=VoseLognormalObject(D5,D6)	=VoseLognormalObjec	=VoseLognormalObj	=VoseLognormalObj
16	Severity (w ded/max)	=VoseDeductObject(B15,B7,B8,TRUE)	=VoseDeductObject(	=VoseDeductObject(D15,D7,D8	=VoseDeductObject(E1	=VoseDeductObject(	=VoseDeductObject
17	Claims (no correlation)	=VoseAggregateMultiMC(B14:G14,B16:G16,{1,0,0,0,0,0;0,1,0,0,0,0;0,0,1,0,0,0;0,0,0,1,0,0;0,0,0,0,1,0;0,0,0,0,0,1})	=VoseAggregateMult	=VoseAggregateMultiMC(B14:G	=VoseAggregateMulti	=VoseAggregateMul	=VoseAggregateMu
18	Total Claims	=SUM(B17:G17)					
19	Admin Costs	=B10*B18					
20	Average Costs	=VoseOutput("average costs (no correlation)","$")+B19/B2					
21	Probability of Profit	=VoseSimProbability(B20,B9)					
22							
23	Claims (0.8 correlation)	=VoseAggregateMultiMC(B14:G14,B16:G16,{1,0.8,0.8,0.8,0.8,0.8;0.8,1,0.8,0.8,0.8,0.8;0.8,0.8,1,0.8,0.8,0.8;0.8,0.8,0.8,1,0.8,0.8;0.8,0.8,0.8,0.8,1,0.8;0.8,0.8,0.8,0.8,0.8	=VoseAggregateMult	=VoseAggregateMultiMC(B14:G	=VoseAggregateMulti	=VoseAggregateMul	=VoseAggregateMu
24	Total Claims	=SUM(B23:G23)					
25	Admin Costs	=B10*B24					
26	Average Costs	=VoseOutput("avearge costs (0.8 correlations)","$")+B25/B2					
27	Probability of Profit	=VoseSimProbability(B26,B9)					

FIGURE 3.27
The multiple damages model.

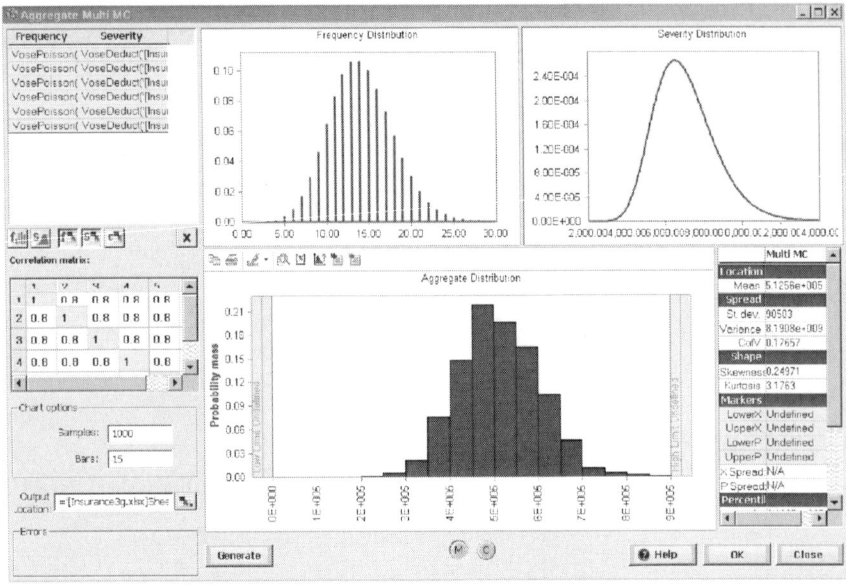

FIGURE 3.28
The AggregateMultiMC window.

The only new feature in this model is the use of the VoseAggregateMultiMC. (We previously used the VoseAggregateMC.) This is a multivariate version of the Aggregate Monte Carlo, which means that it considers several frequency and severity distributions simultaneously. It also permits correlation between the different risks.* We have done two sets of calculations, one with no correlation between these different damage types and the other with an assumed 0.8 correlation between each of the types. The dialog box for the VoseAggregateMultiMC function appears in Figure 3.28.

The frequency and severity functions were entered by pointing to the appropriate cells in the spreadsheet. The "no correlation" calculation uses the default correlation matrix showing zero as the correlation between each pair of damage distributions; the "0.8 correlation" calculation involved changing all of the 0s in the matrix to 0.8. This is an Excel array function, so to enter it in the spreadsheet you must select the appropriate number of cells (in this case either 1 × 6 or 6 × 1 so that there is a cell for each damage type) before pasting it into the spreadsheet.

* We consider correlation more fully in Chapter 7. Correlated risks may result from policies issued in the same region, affected by similar climate, related traffic conditions, etc. There may also be relationships between different types of policies; for example, collision claims and bodily injury claims may be related in terms of frequency, severity, or both.

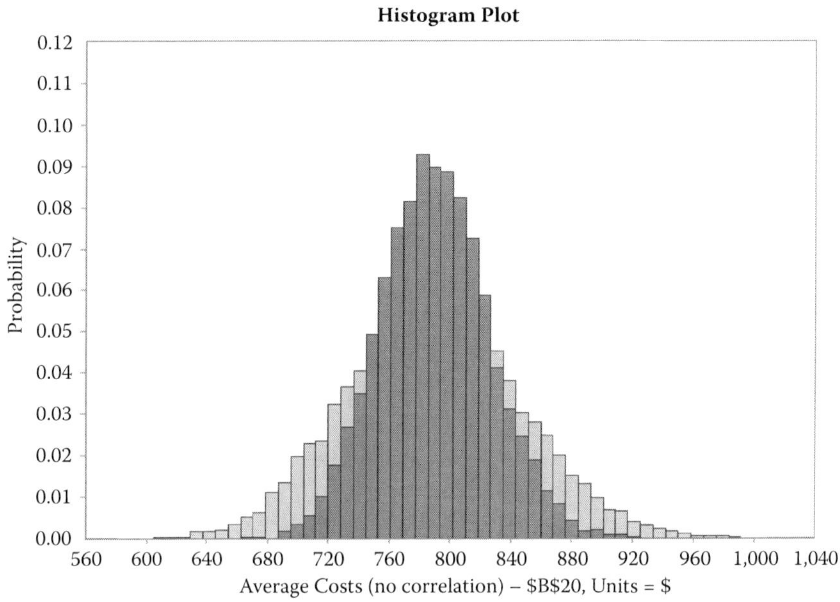

FIGURE 3.29
Results showing the effect of correlation on costs.

Running 10,000 simulations for 1,000 insured drivers* and comparing the outputs with and without correlation between damage types gives the results shown in Figure 3.29. Also, calculations of the "Probability of Profit" in the spreadsheet reveal the probability that the average policy cost is less than the assumed average premium of $800.

The narrower histogram in Figure 3.29 represents the no correlation case: The mean policy cost is not affected by correlation, but the results are clearly riskier when correlation is present. This is because relatively high (or low) claims in one dimension will tend to be matched by relatively high (or low) claims in other dimensions when correlation is present. The wider correlation histogram is also borne out by the probability of profit calculations in the spreadsheet: The probability is 61.77% with no correlation, but only 57.85% with correlation.

This model has incorporated a number of complex considerations. While the frequency and severity data were real, it should be remembered that we made hypothetical assumptions regarding the shapes of the severity distributions, the nonclaim costs of insurers, and the average policy

* With small probabilities, it is a good idea to simulate a large enough number of drivers for claims to occur regularly and to run enough simulations to get reliable results. (Figure 3.29 is based on 10,000 simulations.)

premium, and that we did not consider the effects of different risk classes. Still, this model can be used as a basis for including more realistic and disaggregated assumptions. The next chapter will explore more fully how probability distributions may be chosen. This chapter's exercises will illustrate how frequency/severity modeling can be applied in a number of disparate settings.

3.9 Uncertainty and Variability

A distinction is often made between uncertainty and variability.* Both are modeled with probability distributions, but uncertainty represents lack of knowledge about specific population parameters while variability reflects the fact that random variables (for instance, age or weight) will vary across a population. For example, the number of insurance claims will vary from year to year, and we can represent this variability with probability distributions. However, because the actual average rate at which claims occur is difficult to determine exactly, the rate parameter in the frequency distribution is uncertain, and we can also represent this parameter with a probability distribution.

Thus far, our models have assumed that the rate of claims is known with certainty. Suppose that this is not the case—that, in fact, the rate of claims (assumed to be .07703) is really not known exactly. Suppose that it can be modeled with a Lognormal distribution, with mean .07703 and a standard deviation greater than zero.† Uncertainty about this true parameter value should result in more risk for the insurer. Indeed it does, as we now illustrate.

Figure 3.30 shows the model (Insurance variability3.xlsx; 10,000 policy holders, no deductible, and four cases; no uncertainty about the rate of claims; and standard deviations for the claim frequency equal to 10, 20, and 30% of the mean frequency).‡

Note that the simulated frequency in column C uses Lognormal distributions with the standard deviations given in column B. (Zero standard deviation is the case of no uncertainty.) The model results are shown in Figure 3.31.

* A common distinction is that uncertainty can be reduced through acquiring more information while variability is an attribute of the population that is not reduced with more information. This distinction is not relevant to how we view risk modeling, since both may be modeled with probability distributions.

† A more appropriate distribution to use might be the Gamma distribution, but we will use the more familiar Lognormal distribution in this example. Chapter 4 examines how to choose appropriate distributions.

‡ We construct the full model for the four cases, rather than using a Simtable, because the use of Objects makes the model simple to construct.

number insured	1000		
claim frequency			
mean	0.07703	simulated frequency	rate for Poisson
std deviation	0	=VoseLognormal($B$3,B4)	=C4*$B$1
10% of mean	=0.1*$B$3	=VoseLognormal($B$3,B5)	=C5*$B$1
20% of mean	=0.2*$B$3	=VoseLognormal($B$3,B6)	=C6*$B$1
30% of mean	=0.3*$B$3	=VoseLognormal($B$3,B7)	=C7*$B$1
Severity mean	3321		
Severity std dev	2742		
admin costs	0.1		

Simulation	# claims	Claim amount	Total Claims	Admin costs	Total Cost	Premium	99th percentile of Total Claims
no uncertainty	=VosePoissonObject(D4)	=VoseLognormalObject($B$8,$B$9)	=VoseAggregateMC(B13,C13)	=$B$10*D13	=D13+E13	=VoseOutput(Sheet1!A13)+F13/$B$1	=VoseSimPercentile(D13,0.99)
10% std dev	=VosePoissonObject(D5)	=VoseLognormalObject($B$8,$B$9)	=VoseAggregateMC(B14,C14)	=$B$10*D14	=D14+E14	=VoseOutput(Sheet1!A14)+F14/$B$1	=VoseSimPercentile(D14,0.99)
20% std dev	=VosePoissonObject(D6)	=VoseLognormalObject($B$8,$B$9)	=VoseAggregateMC(B15,C15)	=$B$10*D15	=D15+E15	=VoseOutput(Sheet1!A15)+F15/$B$1	=VoseSimPercentile(D15,0.99)
30% std dev	=VosePoissonObject(D7)	=VoseLognormalObject($B$8,$B$9)	=VoseAggregateMC(B16,C16)	=$B$10*D16	=D16+E16	=VoseOutput(Sheet1!A16)+F16/$B$1	=VoseSimPercentile(D16,0.99)

FIGURE 3.30
Variability/uncertainty model.

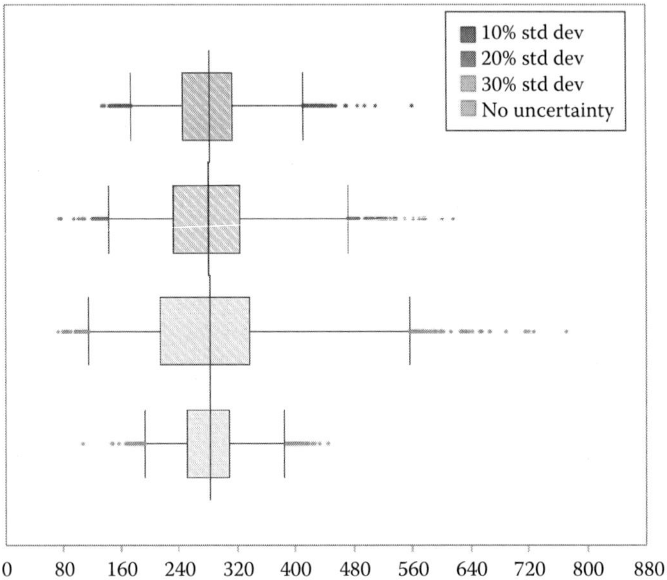

FIGURE 3.31
The effects of variability and uncertainty on average policy cost.

The no uncertainty case shows how variability alone (in claims frequency and severity) affects average policy costs. The total uncertainty in costs per policy increases dramatically as uncertainty about the true frequency probability (the parameter underlying the frequency distribution) increases. The increase in risk for the insurer due to uncertainty about the frequency parameter is graphically evident in Figure 3.32, which shows the 99th percentile of the total claim distribution.

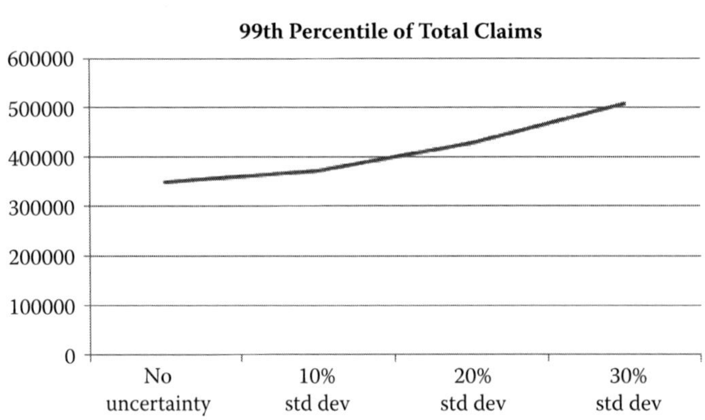

FIGURE 3.32
Impact of uncertainty on the 99th percentile of the total claim distribution.

The 99th percentile for total claim losses rises from around $350,000 for the no uncertainty case to over $506,000 when the standard deviation for the frequency parameter is 30% of its expected value. We leave it as an exercise for the reader to explore how uncertainty about the parameters of the severity distribution might contribute to the insurer's risk. What is clear, however, is that the total uncertainty in risk analysis is magnified greatly when we are also uncertain about the true value of the parameters in the probability distributions and models.

CHAPTER 3
Exercises

3.1 HOSPITAL BED UTILIZATION

The Centers for Disease Control and Prevention report that the average rate of discharge from hospital stays was 1,168.7 per 10,000 people in 2006. The average length of stay was 4.8 days. Consider a medium sized city with a population of 500,000 and five hospitals, each with 155 beds that can be utilized 100% without any downtime between patients. Assume that the rate of daily admissions follows a Poisson distribution and the rate of admissions is evenly spread out over the year. The length of stay is a continuous variable and follows a Lognormal distribution (with the mean given previously and standard deviation of 3 days).

a. What is the probability that this city has adequate hospital beds for the yearly demand for beds? What is the average annual occupancy rate as a percentage of the total beds in the city?
b. Assume that the number of available beds is affected by downtime between patients (and periodic maintenance). Over the course of the year, the number of available beds as a percentage of total beds follows a PERT distribution (minimum = 90%, mode = 93%, maximum = 96%). Recalculate your answer to part (a).
c. How frequently will occupancy rates greater than 80% be observed? Greater than 90%?

3.2 UTILITY COMPANY REPAIR VEHICLES

Consider a utility company with a fleet of 50 repair vehicles. Each day, the number of hours the vehicle will be used follows a Poisson distribution (with an average rate of 4 hours). The number of repair trips taken per vehicle per day follows a Poisson distribution (with an average of one per day). Repair trips require travel that follows a Lognormal distribution (with a mean of 20 miles and a standard deviation of 10 miles). The current truck fleet gets 15 miles per gallon and the average fuel cost is $3.50 per gallon.

a. Estimate the expected annual fuel costs for the repair fleet. Provide a 90% confidence interval for the annual fuel costs.
b. Comment on the current capacity and utilization of the repair fleet.

TABLE 3.1

Insurance Policy Types

Policy Type	Average Premium	Deductible	Claim Frequency	Average Damage/Claim	Percentage Choosing
Regular	2,800	1,400	21.33%	11,433	76.72
Low deductible	3,640	840	27.76%	10,233	21.96
High deductible	1,960	2,520	15.05%	13,600	0.74
Very high deductible	1,920	3,640	11.37%	10,750	0.6

3.3 AUTOMOBILE INSURANCE: MORAL HAZARD AND ADVERSE SELECTION

Data from an automobile insurance company show that claims frequency and size are affected by what type of insurance is chosen by an individual.* The data are from an Israeli insurer for the years 1994–1999. Table 3.1 provides relevant data. Model the claims frequency using a Poisson distribution (to allow for more than one claim per policy holder) and use a Lognormal distribution for the size of claims (with the mean given in Table 3.1 and standard deviation equal to the mean). Assume that 1,000 drivers are insured by the policies, in the proportions indicated in Table 3.1. Also assume that there is no correlation between policy types. Build a simulation model to

a. Estimate the contribution margin (revenue minus direct costs from claims—that is, what is left to cover overhead costs and profits) per insured driver—for each policy type and for the total of all drivers insured by this company
b. See whether the differences in premiums appear to match the differences in expected costs across policies
c. Determine the overall probability that this company will have a positive contribution. What level of contribution do you estimate there is a 95% likelihood of achieving or exceeding?

3.4 CUSTOMER LOYALTY PROGRAMS

Customer loyalty programs are designed to increase revenues from regular customers (and increase their retention).† Suppose that grocery store customer purchasing behavior follows the pattern in

* The data are adapted from Cohen, A. 2005. Asymmetric information and learning: Evidence from the automobile insurance market. *The Review of Economics and Statistics* 87:197–207. We have simplified the data and made some additional assumptions to produce Table 3.2. The numbers are measured in New Israeli shekels.

† For an investigation based on some real data, see Wagner, T., T. Hennig-Thurau, and T. Rudolph. 2009. Does customer demotion jeopardize loyalty? *Journal of Marketing* 73:69–85. That study suggests that, although customers do not increase their purchases when they receive "gold" status, they significantly decrease purchasing behavior when they are demoted from gold status. The latter effect is not surprising, but the former is. We have used hypothetical data for what we believe is the more typical pattern of behavior, where gold status has some positive impact, but demotion does indeed cause purchases to drop.

TABLE 3.2

Customer Loyalty Program Data

Customer Status	Average No. of Annual Transactions	Average Transaction Amount
Normal	39.5	$107.42
Upgraded to gold	43.4	$122.57
Demoted from gold	32.7	$92.21

Table 3.2. Current company policy is to upgrade a customer to "gold" status if yearly purchases exceed $5,000. The customer is demoted from gold status when purchases fall below this amount. Gold status costs the company $400 in yearly benefits. Model the variability in the frequency of purchases using Poisson distributions. Model variability in the level (severity) of purchases using Lognormal distributions (with means given in the table and standard deviations equal to 20% of these means).

a. For a typical year, estimate the probability that a customer in each status will stay in that status, upgrade to gold, or be demoted from gold.
b. Estimate the average gross profit (annual sales minus loyalty program cost) for each status and comment on the profitability of this loyalty scheme for this grocer.
c. Would changing the threshold for gold status to $4,000 improve this loyalty scheme? Assume that customers have the same behavior (Table 3.2) after the change in threshold.

3.5 INTERNET SPAM MARKETING

Spam is a business of very large and very small numbers. Spam volume is estimated to be around 200 billion messages per day, close to 90% of all e-mail traffic.* Suppose that it costs $80 per million spam messages sent and that the conversion rate (the fraction of spam recipients who make a purchase) is 0.0000008 (use a Binomial distribution to represent the number of sales). When sales are made, they follow a Lognormal distribution (with mean and standard deviation each equal to $100). A company decides to send out 1 billion spam e-mails.

a. What is the probability that the spam campaign will be profitable?
b. The conversion rate is actually quite uncertain. Suppose that 0.0000008 is the most likely value, but that experts say that it could be as low as 0.00000001 or as high as 0.000001 (use a PERT distribution to model this).†

---

* Data obtained from www.m86security.com/labs/spam_statistics.asp on October 17, 2010.

† The lower estimate of the sales probability comes from Kanich, C. et al. 2008. Spamalytics: An empirical analysis of spam marketing conversion. Conference on Computer and Communications Security, Arlington, Virginia. Association for Computing Machinery.

TABLE 3.3

Oil Spill Sizes and Frequencies

Spill Size (bbls)	Total Spills/Billion Barrels of Crude Oil Transported
1–49	15
50–999	1.89
1,000–9,999	0.45
10,000–99,999	0.25
100,000–199,999	0.026
>200,000	0.094

3.6 OIL SPILLS

As the world depends on oil that is harder to reach (e.g., from Alaska, deep in the Gulf of Mexico, etc.), the threat of oil spills is ever present. The data in Table 3.3 come from Transport Canada.* For the oil spill sizes, use Lognormal distributions (with means equal to the midpoints for each range and that standard deviations equal to one-third of the difference between each midpoint and the end of that range; for the open-ended category, use 500,000 as the mean and 100,000 as the standard deviation). Model the number of spills of the various sizes using a Poisson distribution. The world transports around 43 million barrels/day of oil via tanker.

a. Develop a forecast for total annual crude oil spill volumes. What is the expected total spill volume? What is the 95th percentile of the distribution (i.e., the volume that there is only a 5% probability of exceeding)? What assumptions have you made?
b. Compare the benefits of alternatively reducing the number of spills of each size by 10%. Does it make more sense to concentrate on reducing the number of large oil spills or small ones?

3.7 SUPERMARKET CHECKOUT

Suppose that customers arrive at a supermarket cash register at the rate of 45 per hour. Each customer takes an average of 2.85 minutes to check out, although this varies. (It follows a Lognormal distribution with a mean of 2.85 minutes and a standard deviation of 3 minutes.)

a. Use Distribution Objects for frequency, severity, and Aggregate distributions to generate a histogram of the number of cashier minutes per hour that are required for customer checkout. How many cashiers would be necessary to have an 85% probability

* Data obtained from www.tc.gc.ca/eng/marinesafety/tp-tp14740-oil-spill-frequency-405.htm on October 16, 2010.

of meeting the demand during any hour? Assume that cashiers are available for all 60 minutes per hour and that there is no time required between customers.

b. Assume that the supermarket employs the number of cashiers you calculated in part (a). Create a histogram of the percentage of their time that is not occupied with customers.
c. Re-create the solution to part (a) *without* the use of Distribution Objects or an Aggregate distribution. Do the results from both methods match?

3.8 SNOWFALL AT A SKI AREA

Data for the years 2001–2009 for the Lake Tahoe area reveal that the average seasonal (October 1 to May 31) snowfall is 183 inches, but ranges from a low of 120 inches to a high of 273 inches. Snowfall is critical to the ski industry. Your job is to provide a 90% confidence interval for the total yearly snowfall expected at Lake Tahoe. Further investigation reveals that it snows on an average of 15.8% of the days during the 243-day snow season (use a Poisson distribution to model the number of days on which it will snow). Further, snowfall amounts on days that it snows closely follow an Exponential distribution (with mean = 4.758).

a. Produce a 90% confidence interval for average seasonal snowfall. What assumptions did you make?
b. What is the probability of getting a season with more snowfall than the highest recorded during the 2001–2009 decade (273 inches)?
c. The 15.8% average of days with snowfall was observed during this time period, but it is only an estimate of the true frequency of snow. Assume that this frequency is only an estimate; the true frequency of snow follows a Lognormal distribution (with mean = 15.8% and standard deviation 1.58). Answer parts (a) and (b) after incorporating this uncertainty into your model.

4

Selecting Distributions

LEARNING OBJECTIVES

- See how distributions may be selected based on theoretical considerations and/or expert opinion.
- Learn how to use the Expert Tools in ModelRisk for selecting distributions and incorporating expert opinions. Apply these methods to company valuation.
- See how distributions may be fit to historical data.
- Learn ways to estimate and incorporate uncertainty about the appropriate distribution that fits the data. Apply these methods to estimation of Value at Risk.
- Understand a variety of classes of distributions commonly used to model particular types of situations with uncertainty or variability.
- Adopt a structured approach to selecting appropriate probability distributions.
- Understand the value of Bayesian Model Averaging (BMA) and how to implement this in a spreadsheet risk analysis model.

4.1 First Introductory Case: Valuation of a Public Company—Using Expert Opinion

In 2010, Apple overtook Microsoft as the world's most valuable technology company, with a total value of over $222 billion.* One way to value a company is through a discounted cash flow analysis. The net cash flows (also known as free cash flows) are projected for a period of time (typically 5–10 years) and a terminal value is added to reflect ongoing cash flows after that time (usually at a fairly low level of sales growth). Discounting is commonly done at the cost of equity, typically calculated using the Capital Asset Pricing Model (CAPM), where the cost of equity includes three components: a riskless interest rate, an equity premium reflecting the riskiness of equity markets overall,

* Apple passes Microsoft as no. 1 in tech. *The New York Times* online, May 26, 2010, accessed on August 22, 2010.

	A	B	C	D	E	F
1		2005	2006	2007	2008	2009
2	Sales	13931	19315	24006	32479	42905
3	Cost of Goods	9888	13717	15852	21334	25683
4	R&D			782	1109	1333
5	SG&A	2393	3145	3745	4870	5482
6	Net Income	1335	1989	3496	4834	8235
7						
8	R&D %			3.00%	3.00%	3.00%
9	SG&A %	17.18%	16.28%	15.60%	14.99%	12.78%
10	Cost %	70.98%	71.02%	66.03%	65.69%	59.86%
11	Sales growth		0.386476	0.242868	0.352953	0.321007

FIGURE 4.1
Apple historical data.

and a company-specific additional risk factor (beta, reflecting the riskiness of a company compared with the market overall). Figure 4.1 shows the most recent 5 years of income statement data for Apple, Inc. (ticker symbol AAPL).

Investment forecasts predict 33.2% sales growth for 2010. In this model, we will base our income statement on future sales projections. We use a cost of goods sold of 65% (comparable to the past 5 years), but with a projected decrease of 1% per year over the next 10 years because technological progress is likely to reduce production costs. Overhead costs (Selling, General, and Administrative expense, SG&A) are projected to be 15% of sales revenue. R&D, important for a technology company, has been stable at 3% of sales and will be forecast to stay at this level. We use a tax rate of 33% (to reflect all relevant taxes), and terminal sales growth (starting in 2020 and continuing indefinitely) is assumed to be 5.5%.* Finally, the cost of equity is estimated to be 15.12%, using a riskless rate of 4%, a market risk premium of 8%, and a company Beta of 1.39 (an average over several sources of estimates).† Based on these inputs, the valuation of Apple comes to just over $222 billion, as shown in Figure 4.2.

While this model provides a reasonable calibration to market beliefs in 2010, it does not reflect the myriad uncertainties that the company faces. Use of expert opinion, combined with Monte Carlo simulation, can provide a sense of the range of plausible values for this company.

* This is quite high; the usual assumption is that the terminal growth rate will equal the rate of growth of the overall economy. However, it is difficult to justify this company's market valuation without incorporating a terminal growth rate that is quite high. Apparently, investors believe that Apple possesses some sustainable competitive advantage that will continue well into the future. It would be more reasonable to assume that the terminal growth rate will gradually decline after year 10 as well. We will not make this adjustment now because it will be considered as part of the simulation in the next section.

† Using CAPM, Cost of Equity = (Market Risk Premium * Beta) + Riskless Rate; that is, (8% * 1.39) + 4% = 15.12%.

	A	B	C	D	E	F	G	H	I	J	K	L	M	N	O	P	Q
1	AAPL																
2	Parameters																
3	Sales growth	33.32%					33.32%	30.54%	27.76%	24.97%	22.19%	19.41%	16.63%	13.85%	11.06%	8.28%	5.50%
4	COGS	65%															
5	annual cost decrease	1%															
6	R&D %	3%															
7	SG&A%	15%															
8	terminal growth	5.50%															
9	Beta	1.39				cost of equity	15.120%										
10	riskless	4.00%															
11	risk premium	8.00%															
12	tax rate	33%															
13																	
14		2005	2006	2007	2008	2009	2010	2011	2012	2013	2014	2015	2016	2017	2018	2019	2020
15	Sales	13,931	19,315	24,006	32,479	42,905	57200.95	74668.971	95394.09	119217.8	145674.6	173950.1	202874.5	230964.5	256518.4	277763.3	293040.2
16	Cost of Goods	9,888	13,717	15,852	21,334	25,683	37180.61	48049.483	60772.24	75190	90957.4	107526.1	124151.5	139928.1	153855.6	164932	172263.2
17	R&D			782	1,109	1,333	1716.028	2240.0691	2861.823	3576.534	4370.239	5218.502	6086.235	6928.935	7695.552	8332.898	8791.207
18	SG&A	2,393	3,145	3,745	4,870	5,482	8580.142	11200.346	14309.11	17882.67	21851.19	26092.51	30431.17	34644.67	38477.76	41664.49	43956.04
19	Net Income	1,335	1,989	3,496	4,834	8,235	9724.161	13179.073	17450.92	22568.61	28495.79	35112.95	42205.58	49462.78	56489.44	62833.88	68029.77
20	post-tax income						6515.188	8829.9792	11692.12	15120.97	19092.18	23525.67	28277.74	33140.07	37847.93	42098.7	45579.95
21	R&D %			3%	3%	3%											
22	SG&A %	17.18%	16.28%	15.60%	14.99%	12.78%											
23	Cost %	70.98%	71.02%	66.03%	65.69%	59.86%											
24	Sales growth		38.65%	24.29%	35.30%	32.10%											
25																	
26	Objectives																
27	PV net income (10 years)																
28	$100,084.30																
29	Terminal Value																
30	$499,863.27																
31	PV terminal value																
32	$122,276.62																
33	Total Value																
34	$222,360.92																
35	Terminal Value %																
36	54.99%																

FIGURE 4.2
Apple valuation spreadsheet.

4.2 Modeling Expert Opinion in the Valuation Model

The people who develop and run risk analysis models are often not the people with the best insight concerning the range and probability of critical values used in that analysis. Further, any one individual is likely to be wrong in his or her predictions, while use of several divergent opinions may produce more reliable estimates.* For the Apple, Inc. valuation, we have identified four key uncertain variables in our analysis: the Cost of Goods Sold (COGS) percentage, the R&D percentage, the Terminal Growth Rate, and Beta (which directly impacts the discount rate).

As Apple's revenues increase, we do not know whether the company will be able to sustain cost decreases or the extent to which it will need to continue to invest in R&D. Many valuation exercises assume that the Terminal Growth Rate will equal the expected growth rate of the economy, but Apple's valuation appears to suggest that the market believes that Apple has sustainable competitive advantages that will enable it to continue to grow at a larger rate. These uncertainties also make Apple's stock riskier than the market overall, and this will impact its cost of capital. While we may have our individual opinions on these key risk factors, we have instead turned to valuation experts to select appropriate probability distributions for each factor.

We have solicited the minimum, most likely, and maximum values, for each risk factor, from three valuation experts.† ModelRisk has an Expert tool (shown in Figure 4.3) that can assist with choosing an appropriate distribution for these parameters. The Expert tool has two tabs, Selector and Shaper. The Selector tab is shown in Figure 4.4, with a set of values entered for the minimum, mode, and maximum values (after pressing the Calculate list button).

We see that there are four potential distributions (that use the minimum, mode, and maximum values of zero, one, and two, respectively) from which to choose in modeling an expert's opinion, along with graphs and statistics about each possible choice. If we had entered alternative parameters, we would have been given different choices of possible distributions, so it is a good idea to investigate how the distribution choices vary with the

* In fact, the use of prediction markets is based on the notion that the "wisdom of crowds" can produce more accurate assessments of uncertain events than what would result from any particular expert's opinion. For examples, see Spann, M. and B. Skiera. 2008. Sports forecasting: A comparison of the forecast accuracy of prediction markets, betting odds and tipsters. *Journal of Forecasting* 28:1. Another example can be found in Kaufman-Scarborough, C. et al. 2010. Improving the crystal ball: Harnessing consumer input to create retail prediction markets. *Journal of Research in Interactive Marketing* 4:1. We discuss this further later in this chapter.

† We thank Jason Pierce, Brenda Schultz, and Mary Wladkowski for their assistance in this case.

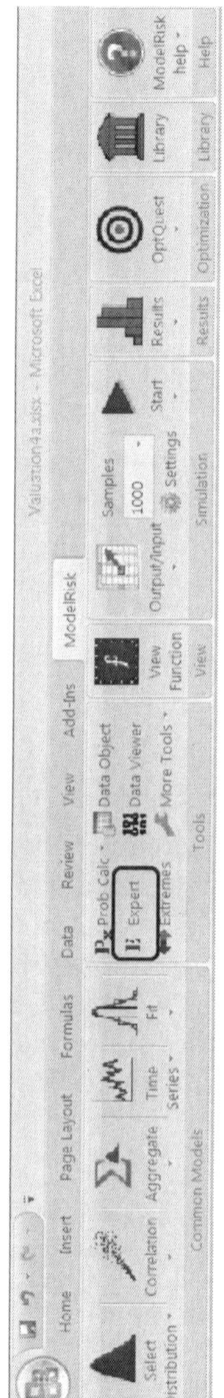

FIGURE 4.3
ModelRisk Expert tool.

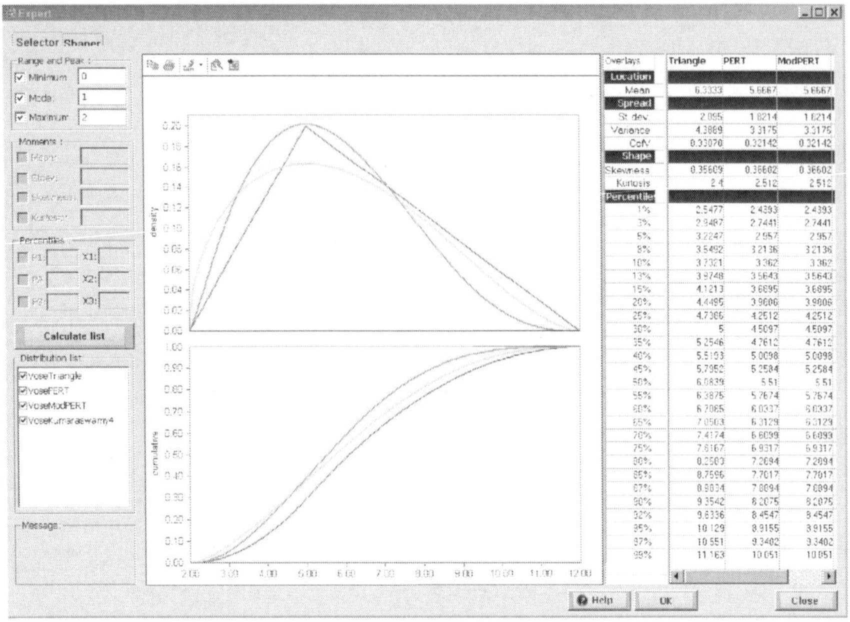

FIGURE 4.4
Expert Selector options.

parameters chosen *before* soliciting the expert opinions since these choices will limit the distributions available.* We will use the PERT distribution here.†

We now need to combine the three PERT distributions obtained from our experts. ModelRisk has a Combine option under Select Distribution on the ModelRisk toolbar. This dialog allows us to enter our three PERT distributions (as Objects) as well as to assign different relative weights to each expert's opinion.‡ We have chosen to assign 40% weights to the two experts who do valuation professionally and a 20% weight to the academically trained expert, but this can be easily changed. Entering this information produces the combined distribution shown in Figure 4.5.

---

* We will not show the Shaper tab here, but it is easy to use and permits a distribution to be built using a number of point estimates from the probability distribution. It may be used interactively with subject matter experts to derive a distribution that best matches their opinion.

† While the Triangle distribution is often used by novice modelers and is easy to understand, it is very sensitive to the exact values chosen for the minimum and maximum estimates, and these are often not as reliably chosen by experts compared to the most likely value. The PERT is not impacted as dramatically by changes in the minimum and maximum values and is therefore preferred.

‡ The distributions do not need to be the same type. Using alternative Monte Carlo simulation software (i.e., not ModelRisk) combining the expert distributions can also be done, but would need to be done manually.

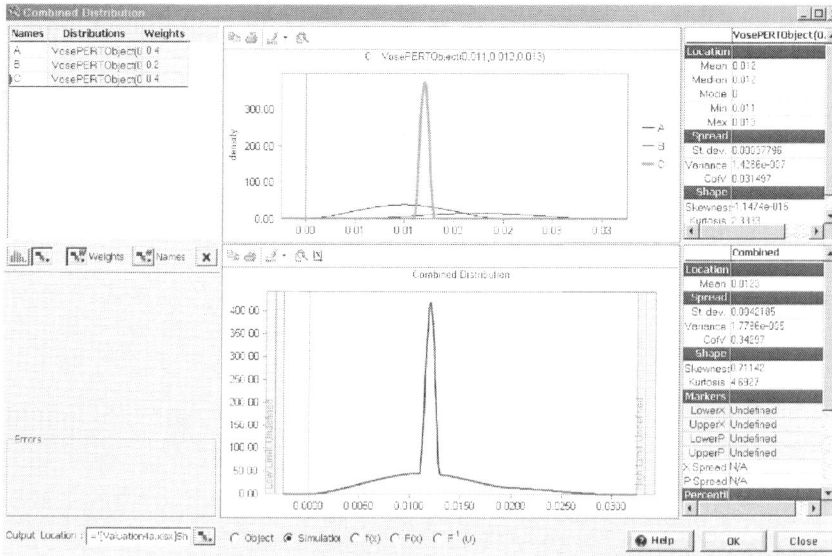

FIGURE 4.5
Combined distribution.

We have done this for each of the four key parameters and marked them as ModelRisk inputs. Running the simulation provides the valuation results shown in Figure 4.6 and the tornado chart in Figure 4.7.

The mean value for Apple is $237 billion, with a 90% confidence interval of ($174 billion, $341 billion). The terminal value averages 55% of the total value of the company. A reasonable further step would be to present these results to the experts and ask them to reassess their choices for Beta, since Beta represents the relative risk of this stock to other equities, and the simulation results provide estimates of the uncertainty surrounding the valuation of this particular stock. (Note: the simulation does not provide estimates on the *relative* risk; this would require comparing simulation valuation models for different companies.)

The Tornado chart reveals that R&D percentage has almost no effect and the annual cost decrease only has a small effect on the uncertainty in future company cash flows. It is perhaps not surprising to find that the Terminal Growth Rate and Beta are the most significant determinants of uncertainty. Over half of Apple's market value is embodied in the value beyond 10 years, and the Terminal Growth Rate is a fundamental determinant of this. Beta will determine the weight that is placed on present versus future earnings, and Apple's expected value of $237 billion is much greater than its 2009 after-tax earnings of around $5.5 billion. (Even 20 years of annual earnings at this level fall far short of the market valuation of this company; its value is dependent on further earnings growth.)

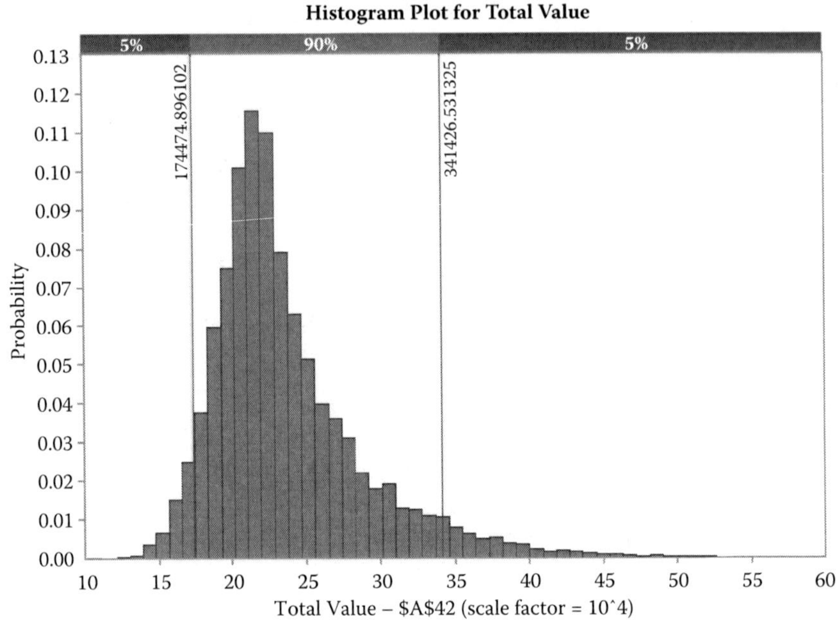

FIGURE 4.6
Valuation simulation results.

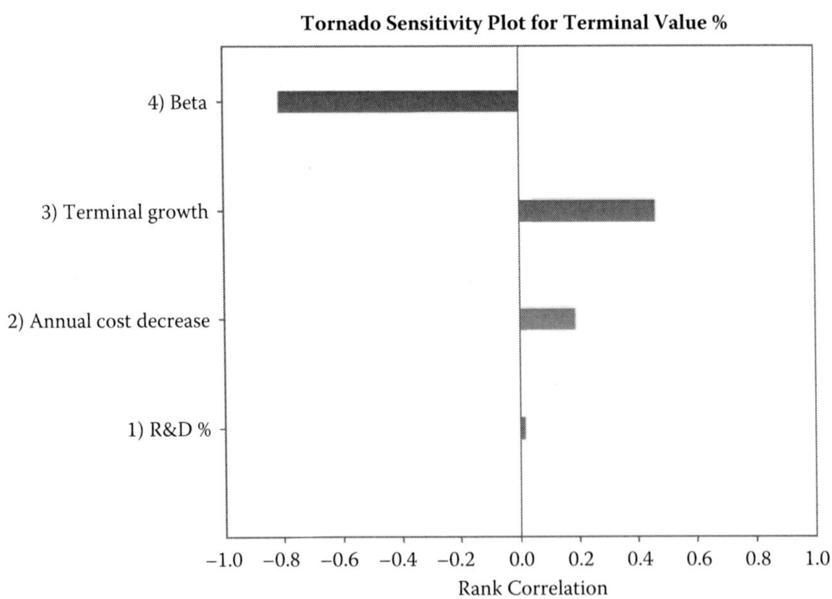

FIGURE 4.7
Valuation tornado chart.

4.3 Second Introductory Case: Value at Risk—Fitting Distributions to Data

The credit and financial crisis of 2007–2009 has focused attention on current practices of risk measurement and management in financial markets. Regulatory agencies throughout the world have adopted Value at Risk (VaR)* and stress testing† techniques to measure the exposure faced by financial institutions. For example, the Basel Committee on Banking Supervision is revisiting the risk guidelines established by the Market Risk Amendment (updated in 2006). The first formal system adopted by the Basel Committee required certain institutions to calculate a 99th percentile, left-tail confidence interval in deriving the VaR measure. Management is mostly free to choose among various methods for estimating this, including historic simulations and Monte Carlo simulations. The chosen methods are back-tested against historical data to track the frequency with which actual losses exceed those that are expected based on the confidence interval. Capital adequacy is determined using this estimation of VaR.

Stress testing takes VaR one step further by asking about the potential size of losses that can occur outside this confidence interval. So, the basic task for both VaR and stress testing is to estimate the probability of losses of various sizes for a given portfolio of assets. Monte Carlo simulation is an excellent tool for establishing these estimates.

4.4 Distribution Fitting for VaR, Parameter Uncertainty, and Model Uncertainty

We explore estimating value at risk for a hypothetical portfolio of $1,000 invested in a single stock. Daily data are available in the spreadsheet VaR.xlsx for 1 year of daily stock prices.‡ Figure 4.8 shows the daily closing price and day-to-day change in the price of this stock. The daily change could be calculated as a percentage return by dividing the difference in the daily closing prices by the beginning price (and multiplying by 100). Instead, we use the continuous version, the natural logarithm of the ratio of successive daily

* There are a number of definitions for Value at Risk. In general, VaR can be thought of as measuring the "maximum" expected future loss from a portfolio (in dollars or as a percentage) at a specific confidence level and within a specific period of time.

† Stress testing also has a number of definitions in the context of a financial institution. In general, stress testing is the process of examining the behavior of a financial instrument, portfolio, or institution under extreme and crisis scenarios (liquidity, inflation, interest rate shocks, economic depression, etc.).

‡ The stock is AIG and the time period is 1/26/2010–1/25/2011.

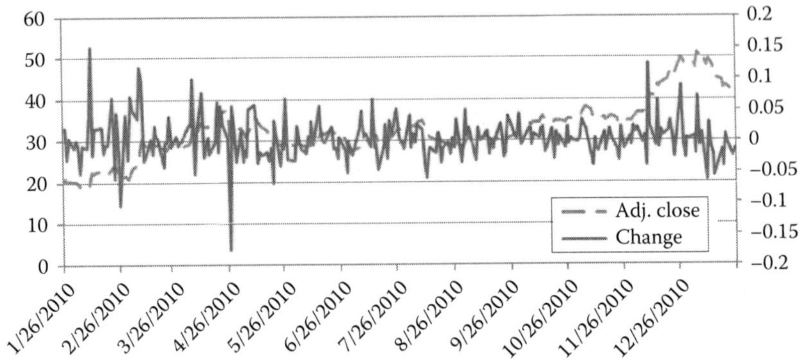

FIGURE 4.8
AIG historical data.

closing prices—often called the *log return*. These two calculations provide nearly identical results based on the fact that $\ln(1 + x) \approx x$ for small values of x.

Time series patterns, such as those shown in Figure 4.8, are typical for equities: The prices follow a meandering pattern, while the period-to-period changes randomly fluctuate around the long-term trend.* From the historic data, the fifth percentile of the distribution is –4.09% (meaning that 5% of the daily changes were a decrease of at least 4.09%) and the first percentile is –6.68%.

We will estimate the 1-day VaR for this stock holding. One approach is to use these historic returns to estimate the 95% VaR and 99% VaR. (The 95% VaR means that there is a 95% chance that the losses in a portfolio will not exceed the estimated amount within a certain time frame—for example, a month, or a day in this case; this corresponds to the fifth percentile of the price change distribution in that specific period.) The use of historically observed losses, however, relies entirely on what has been observed during the time period for which data have been collected. An alternative VaR estimation technique involves fitting a distribution to the historical price changes and then simulating what the losses might look like over a specific period of time (1 day in this example). VaR.xlsx contains a number of models (of increasing complexity) that accomplish this.

The simplest approach is to fit a distribution to the 5 years of data. On the ModelRisk toolbar, click the drop-down menu under Fit, shown in Figure 4.9, and choose the first option, Fit Distribution. Enter the range for

* Of course, if there was any discernable trend in the changes that is predictable for the future, investors would have acted on these and the predictability of the trend would disappear. At least this is the assumption of the *Efficient Market Hypothesis (EMH)*. Even under the EMH, the volatility of the time series of changes may still exhibit predictable patterns, but ignore this in the present chapter. Chapter 6, dealing with time series (forecasting), will permit us to model volatility patterns in the data.

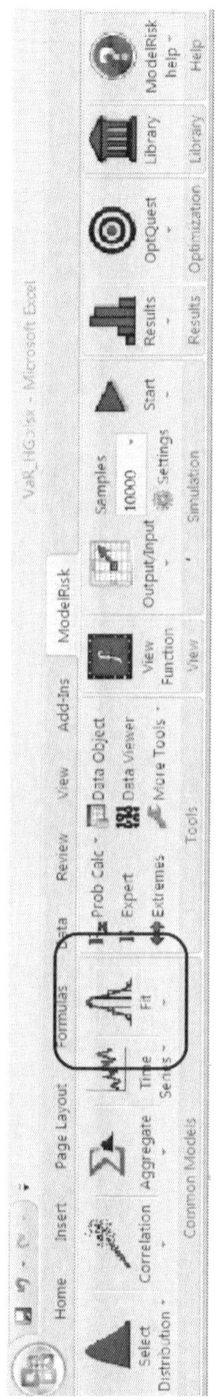

FIGURE 4.9
Fit Distribution button.

the daily price changes (C3:C254) and click the Add button and add all continuous distributions (by highlighting them and clicking the right-arrow button). ModelRisk will not allow some of the distributions to be selected, but do not worry because these distributions cannot be fit to these particular data. (For example, the presence of negative values precludes some distributions, such as the Lognormal.) ModelRisk will fit each appropriate distribution to the data and compute several values, called Information Criteria, to describe how well each type of distribution fits the data. These Information Criteria (SIC, AIC, and HQIC) are based on estimating the likelihood that the observed data could have come from each postulated distribution. Higher numbers (lower absolute values if they are negative) indicate better fitting distributions, and ModelRisk lists the distributions from best fit to worst. (You can choose which criterion to use to rank the distributions because they might rank differently for each criterion.) Figure 4.10 shows the completed Fit Distribution window.

The Laplace distribution fits best (according to the SIC criterion), although several distributions are close. You can click on each distribution in the list to see how it fits the actual data and you should certainly do so. Regardless of the values of the information criteria, you should always visually check to ensure that the distribution appears to be a reasonable description of the data. Clicking the Insert in Worksheet button will provide several options for placing the distribution into the spreadsheet for simulation. Choosing

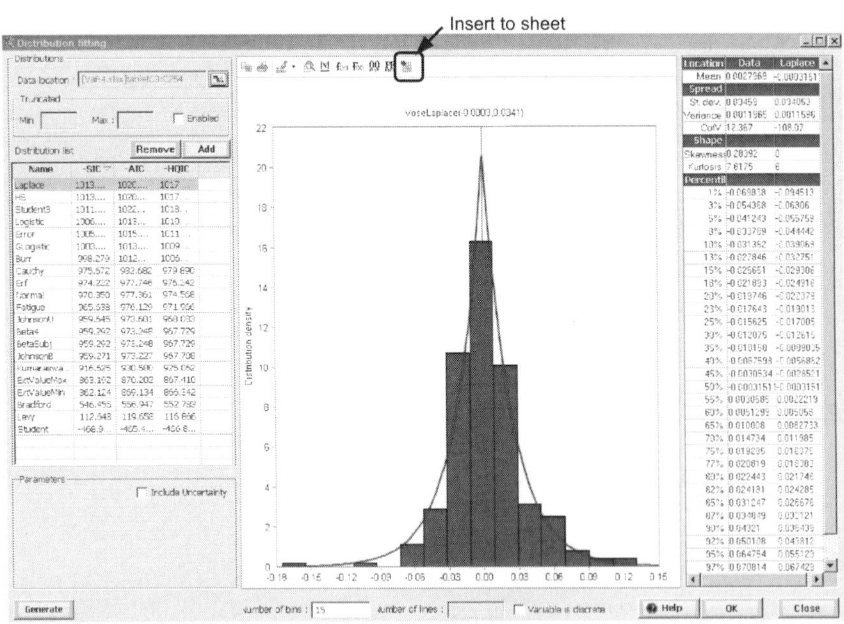

FIGURE 4.10
Fitted distributions for daily price changes.

the Simulate option directly places a simulation from the chosen distribution into the spreadsheet; choosing Object will insert the distribution as an Object, which can be simulated or further manipulated.

We could complete a VaR analysis at this point, using the Laplace distribution, but let us compare it to a couple of additional modeling options. The best fitting Laplace distribution has a particular mathematical form and ModelRisk has estimated the parameters of this function that best fit the data. However, even though we obtained the distribution with the parameters that best fit the data, we will have uncertainty about the true values of these parameters. If we think of the observed data as a random sample from an underlying process (which follows a Laplace distribution), then we might get somewhat different parameters based on a different set of random values. In both cases, the sets of random samples came from the same underlying process, but the fitted distribution for each set will produce different parameters for the Laplace distribution. If we click the Include Uncertainty box in the Distribution Fitting window, we get several potential Laplace distributions that might have resulted from fitting a random sample of data (from the same underlying process) to a Laplace distribution. Figure 4.11 shows five such distributions. (You can change the number shown by changing the Number of Lines entry in the window.)

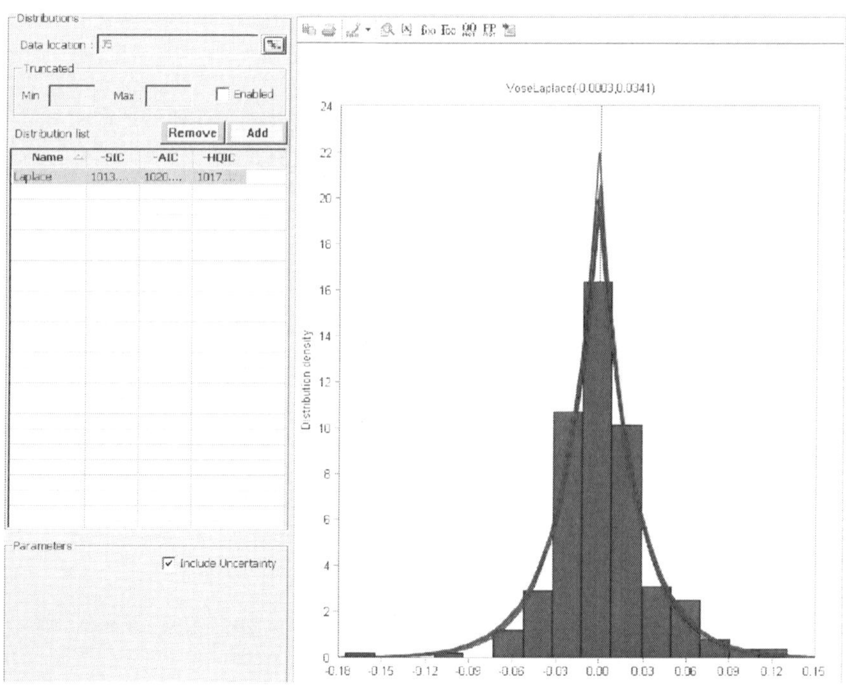

FIGURE 4.11
Parameter uncertainty.

Each of the curves is a Laplace distribution, but with different parameters. We will refer to this as modeling *parameter uncertainty*. Including this uncertainty in fitting a distribution to data is sometimes called a *Second Order distribution fit*, to be contrasted with the *deterministic* (or *First Order*) *fit*, which uses the single best fitting set of parameters (without the Include Uncertainty box checked). If you insert this Object into the spreadsheet, you will see that the First Order object and Second Order object are distinguished by the presence of TRUE at the end of the ModelRisk function. If we use the Second Order Object to model VaR, we will be simulating our uncertainty about the true parameters of the Laplace distribution, in addition to the variability inherent in Monte Carlo simulation of a single Laplace distribution.

We can go one (big) step further. Examining Figure 4.10 shows that a number of distributions provide similar fits to the actual data. The five choices (Laplace, HS, Student3, Error, and GLogistic distributions) all appear close, in terms of both the information criteria and the visual fit to the data. Indeed, the HS distribution ranks as the best fitting distribution by two of the Information Criteria. While we could rely on the Laplace as the best fitting distribution, we do not really know the correct distribution to represent the underlying random process. ModelRisk provides a way to represent this *model uncertainty* and simulate the reality that any of these distributions might be the correct one to use. Bayesian Model Averaging (BMA) is a technique that permits a number of distributions to be used in a Monte Carlo simulation, weighting them according to how well they fit the actual data. Each of these distributions can include parameter uncertainty, but BMA also simulates our uncertainty about which type of distribution best represents the underlying uncertainty. The BMA function requires the use of Objects, and VaR.xlsx contains all three approaches to modeling VaR: the best fitting First Order distribution, the inclusion of parameter uncertainty, and model uncertainty. The portion of the spreadsheet showing the three modeling approaches is shown in Figure 4.12 (with the formulas displayed).

For convenience, we have placed a Data Object in cell J5; this permits us to refer to this single cell rather than needing to reference the entire range containing the historic data. The VoseLaplaceFitObject fits the best First Order Laplace distribution to these data (cell J8), and cells J10 and K10 will provide the fifth and first percentiles of this distribution. The fifth and first percentiles can be obtained directly from the fitted First Order distribution. Parameter uncertainty (using the Second Order fit) is modeled in J13, distinguished by the argument TRUE in the VoseLaplaceFitObject function. We then simulate this Object and place the mean (using VoseSimMean) for the fifth and first percentiles in the spreadsheet.

Model uncertainty, using BMA, requires that we list the potential distributions—each as a FitObject function—using the original Data Object and using TRUE to include parameter uncertainty in each distribution. Cells L19:L23 are required by the BMA function; they represent prior beliefs about the relative weights to be assigned to each potential distribution. We have

	I	J	K	L
1		Historic 5%	Historic 1%	
2		=PERCENTILE(C2:C254,5%)	=PERCENTILE(C3:C254,1%)	
3				
4		**Data Object**		
5		=VoseDataObject(C3:C254)		
6				
7	I	**Static Fit**		
8		=VoseLaplaceFitObject(J5)		
9		static 5%	static 1%	
10		**=VoseSimulate(J8,5%)**	**=VoseOutput("1st Order 1%")+VoseSimulate(J8,1%)**	
11				
12	II	**Parameter Uncertainty**		
13		=VoseLaplaceFitObject(J5,TRUE)		
14		**=VoseSimulate(J13,5%)**	**=VoseOutput("2nd Order 1%")+VoseSimulate(J13,1%)**	
15		=VoseSimMean(J14)	=VoseSimMean(K14)	
16				
17	III	**Model Uncertainty**		
18		Distributions		Priors
19		Laplace	=VoseLaplaceFitObject($J$5,TRUE)	1
20		HS	=VoseHSFitObject($J$5,TRUE)	1
21		Student3	=VoseStudent3FitObject($J$5,TRUE)	1
22		Error	=VoseErrorFitObject($J$5,TRUE)	1
23		Glogistic	=VoseGLogisticFitObject($J$5,TRUE)	1
24				
25				
26				
27		=VoseBMAObject(K19:K23,L19:L23)		
28		**=VoseSimulate(J27,5%)**	**=VoseOutput("BMA 1%")+VoseSimulate(J27,1%)**	
29		=VoseSimMean(J28)	=VoseSimMean(K28)	

FIGURE 4.12
Three VaR models: formulas.

simply assigned equal prior weights to each distribution. The VoseBMA function will begin with these prior weights but will actually weigh each distribution according to the likelihood that the data came from the specific distribution (expressed by a parameter called the Log Likelihood, indicating how well each distribution fits the data). The BMA Object appears in cell J27, and the fifth and first percentiles are simulated below it. Figure 4.13 shows what this section of the spreadsheet looks like without the formulas displayed.

Note that the cells containing the distributions with parameter uncertainty do not actually display TRUE, but rather display the fitted distribution's parameters. Pressing the F9 key will cause these numbers to change (reflecting that uncertainty has been set to TRUE). Also, note that the VoseSimMean functions display "No simulation results" until a simulation has been run.

To complete the VaR estimations, we append 1 day to our historic data and simulate how our initial $1,000 investment changes in the forecast day for each of our three methods (First Order uncertainty, Second Order uncertainty, and model uncertainty). This section of the spreadsheet (with formulas) is shown in Figure 4.14.

Columns J, K, and L simply simulate the daily price change from the models in Figure 4.7, and these are used to compute the investment's value after 1 day. To compute the 1-day VaR, we simulate the first and fifth percentiles of the portfolio values after the simulated day and subtract these amounts from the initial capital. We also compute the size of the loss, the conditional

	I	J	K	L
1		Historic 5%	Historic 1%	
2		-4.09%	-6.68%	
3				
4		**Data Object**		
5		VoseDataObject('[VaR4new.xlsx]table'!C3:C254)		
6				
7	I	**Static Fit**		
8		VoseLaplace(-0.000315,0.034053)		
9		static 5%	static 1%	
10		**-5.58%**	**-9.45%**	
11				
12	II	**Parameter Uncertainty**		
13		VoseLaplace(-0.000177,0.038034)		
14		**-6.21%**	**-10.54%**	
15		No simulation results	No simulation results	
16				
17	III	**Model Uncertainty**		
18		Distributions		Priors
19		Laplace	VoseLaplace(0.001185,0.032452)	1
20		HS	VoseHS(-0.002059,0.031739)	1
21		Student3	VoseStudent3(0.000874,0.035486,4.000000)	1
22		Error	VoseError(0.004525,0.033269,0.963247)	1
23		Glogistic	VoseGLogistic(-0.003059,0.018528,1.173086)	1
24				
25				
26				
27		VoseBMA('[VaR4new.xlsx]table'!K19:K23,'[VaR4new.xlsx]table'!L19:L23)		
28		**-5.06%**	**-9.15%**	
29		No simulation results	No simulation results	

FIGURE 4.13
Three VaR models: cell results.

VaR (cVaR), and the two worst-outcome days that might occur over the next year. (This type of information is commonly used in stress testing, and the VoseKthSmallest function provides these estimates.) The conditional VaR shows the expected loss, conditional on a loss greater than the VaR threshold. In other words, it is the mean of the outcomes that is worse than the VaR estimate. Without the formulas displayed, this section of the spreadsheet appears in Figure 4.15.

Notice that many cells will display "No simulation results" until after the simulation is run, and the cells that use formulas based on those cells display the Excel #VALUE! error. All of these will be replaced by numbers after the simulation is run. We have also calculated the worst and second worst daily price changes expected over the next year (cells E173:G173 and E174:G174 for the three models), using the VoseKthSmallest function. We are now ready to perform the simulations. The numerical results for 10,000 simulations are shown in Figure 4.16.

The VaR estimates increase when parameter uncertainty is included (the second column compared with the first) but then decrease when model uncertainty is included (the third column). Including parameter uncertainty typically introduces more total uncertainty compared to using a First Order fitted distribution. The effects of including model uncertainty are not as predictable, however. This is due to the fact that the best fitting distribution

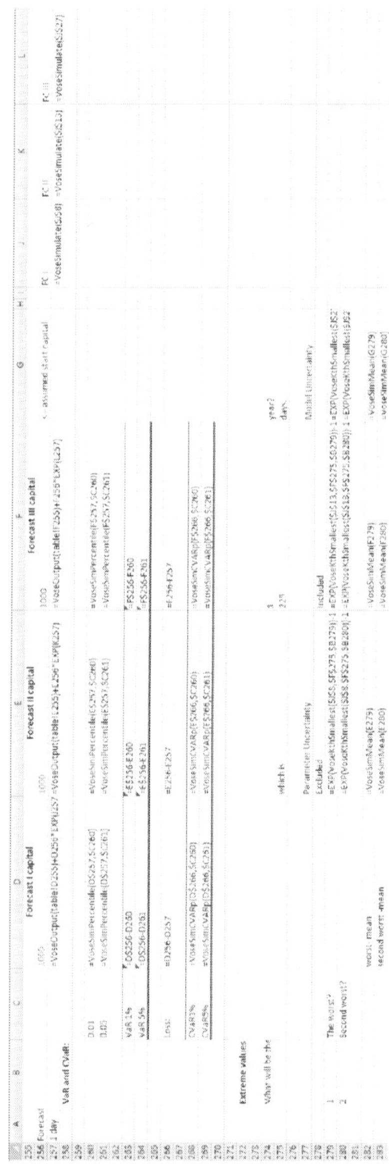

FIGURE 4.14
VaR calculation formulas for a simulated day.

	A	B	C	D	E	F	G	H	I	J	K	L
255				**Forecast I capital**	**Forecast II capital**	**Forecast III capital**						
256	Forecast			$1,000	$1,000	$1,000	<-- assumed start capital			FC I	FC II	FC III
257	1 day			$974	$1,022	$979				-0.0261	0.0215591	-0.021307
258		**VaR and CVaR:**										
259												
260			1%	No simulation results	No simulation results	No simulation results						
261			5%	No simulation results	No simulation results	No simulation results						
262												
263			VaR 1%	#VALUE!	#VALUE!	#VALUE!						
264			VaR 5%	#VALUE!	#VALUE!	#VALUE!						
265												
266			Loss:	$26	($22)	$21						
267												
268			CVaR1%	No simulation results	No simulation results	No simulation results						
269			CVaR5%	No simulation results	No simulation results	No simulation results						
270												
271												
272		**Extreme values:**										
273												
274		What will be the worst return in the next				1	year?					
275					which is	225	days.					
276												
277					Parameter Uncertainty		Model Uncertainty					
278					Excluded	Included						
279			1 The worst?		-11.25%	-14.46%	-8.84%					
280			2 Second worst?		-7.76%	-9.87%	-12.10%					
281												
282				worst- mean	No simulation results	No simulation results	No simulation results					
283				second worst -mean	No simulation results	No simulation results	No simulation results					

FIGURE 4.15
VaR calculation formula cell results.

VaR and CVaR:				
1%	$911	$909	$913	
5%	$947	$945	$950	
VaR 1%	$89.06	$91.37	$86.75	
VaR 5%	$53.02	$54.60	$49.89	
Loss:	($27)	$2	$29	
CVaR1%	$ 110	$ 112	$ 122	
CVaR5%	$ 75	$ 77	$ 75	
Extreme values:				
What will be the worst return in the next			1	year?
		which is	225	days.
		Parameter Uncertainty		Model Uncertainty
		Excluded	Included	
1 The worst?		-11.83%	-7.94%	-10.95%
2 Second worst?		-9.09%	-8.21%	-7.87%
worst- mean		-11.97%	-11.92%	-11.81%
second worst -mean		-9.87%	-9.80%	-9.25%

FIGURE 4.16
VaR numerical results.

type may turn out to be a particularly variable one (i.e., with fatter tails), while some of the other potential distributions could have less variance. So, while model uncertainty accounts for more uncertainty in the simulation (uncertainty about which distribution is the best fitting one), the different shapes of these alternative distributions could result in a decrease in VaR estimates. In the present case, VaR decreases with model uncertainty.

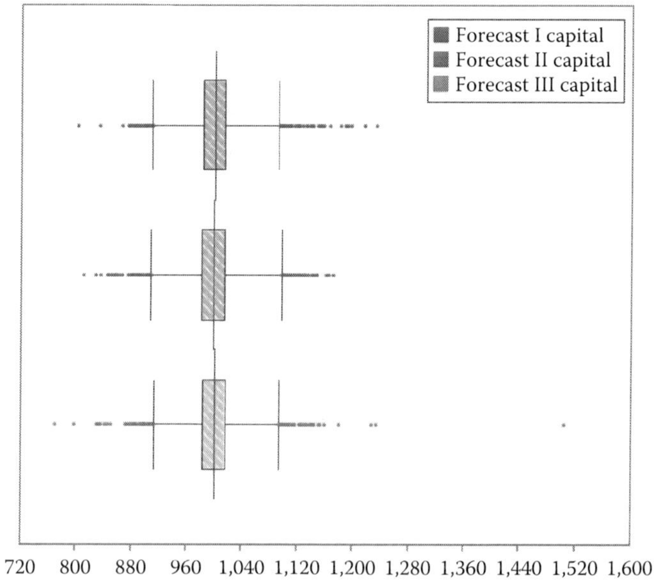

FIGURE 4.17
Distributions of capital after a simulated day.

Most notable, however, is that the numbers only change slightly across the three models. Figure 4.17 underscores this by showing box plots for ending capital under First Order distribution fitting, Second Order uncertainty, and model uncertainty.

The box shows the middle 50% of the simulations and the lines extending on each side of the box (called "whiskers"—hence the name "box and whisker plot") show approximately the first and last quartiles of the simulations. The ends of the line segments (the "whiskers") can be manually set and are here set to show the 1st and 99th percentiles of the distribution. Outliers are shown as isolated points beyond these lines. The standard deviation actually increases slightly as more uncertainty is included in the model, but the distributions are remarkably close.

The real difference between these methods only becomes apparent if we examine the first percentiles of the distributions of daily simulated returns. These are shown in Figure 4.18.

Recall that the first percentile of the historical data was a loss of 6.68%. First Order uncertainty shows a loss of 9.13%, but there is no uncertainty about this. Once a static distribution has been fit to the data, the first percentile can be computed and is no longer uncertain: 1% of the time, losses should exceed 9.13%. It is uncertain whether we will experience such a loss, but the first percentile itself is no longer uncertain. But with Second Order uncertainty and model uncertainty, the first percentile is uncertain. In this case, Second Order uncertainty is larger than model uncertainty, but both

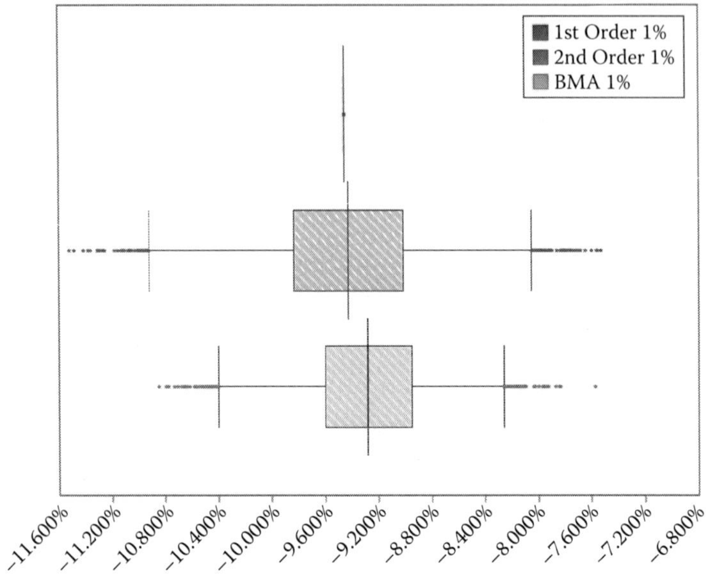

FIGURE 4.18
Distributions of the first percentiles of daily returns.

show that a range of values is possible for the first percentile. The VaR only changes slightly because it is based on the means of these distributions, which do not vary by much. Notice that the first percentile of the Second Order uncertainty distribution (for the first percentile of daily returns) is –10.9% and the middle 50% of the distribution is in the range of (–9.8%, –9%).

VaR estimates are based on the idea that the VaR is a single number, rather than a range of possible numbers. It is not clear that the mean of the distribution of 1% VaR values is the appropriate benchmark for measuring value at risk. Using the first percentile of the 1% VaR is probably too restrictive (for judging capital adequacy). But using the mean of the uncertain 1% VaR may not adequately protect against risk. This is an unresolved issue with the use of VaR.*

We used the VaR example to illustrate how parameter and model uncertainty can be included in a simulation model. While the "average VaR" was not much affected by including the additional uncertainty, our primary finding is that VaR is not a number, but actually a distribution.

* An even more uncertain view of VaR can be found in Hsu, J. C. and V. Kalesnik. 2009. Risk-managing the uncertainty in VaR model parameters. Chapter 18 in Gregoriou, G. *The VaR Implementation Handbook*. New York: McGraw–Hill. These authors suggest the use of a variety of expert judgments in representing uncertainty about return distribution parameters. This would be an excellent application for use of the Expert Tools (Section 4.2) in a VaR model. Instead of uncertainty about the parameters of the distribution being estimated from the historical data, it could vary according to expert opinion. (Similarly, the distribution itself could also be subject to differing expert opinions.)

We caution readers that these comparisons are likely to vary across data sets. Comparisons of the models will depend on the volatility of the data, the actual form of the best fitting distributions, and the amount of data available. In particular, as more data are available, the differences between these methodologies will decrease. It also seems that parameter uncertainty will have more profound impacts on skewed distributions than on symmetric ones.

While ModelRisk includes powerful techniques for fitting distributions to data and the ability to include and quantify the uncertainty about the fitted distributions, we end with two caveats. First, it is important always to look at the distributions you are fitting and not rely solely on the Information Criteria measures of fit. You may be especially interested in a particular part of the distribution (such as the left-hand tail), so it may be the distribution that fits this part of the data best that is most relevant; however, that may not be the distribution that fits best overall. Second, you should make sure the distribution that you are fitting makes sense for the problem you are modeling. Each distribution possesses properties (such as the extent to which tail probabilities can occur) that may not be appropriate for your problem. **Quantitative measures of fit cannot substitute for the use of judgment in modeling.**

4.5 Commonly Used Discrete Distributions

There are around two dozen discrete and four dozen continuous probability distributions in ModelRisk. We will not provide an exhaustive description of each—full documentation, including mathematic details and example models, is contained within the ModelRisk help file.* Many of these distributions are rarely used on theoretical grounds; they may, however, fit a given set of data well. Others have very specialized applications. Our intent in this section (for discrete distributions) and the next section (for continuous distributions) is to review the most commonly used distributions and circumstances under which they are potential candidates for representing uncertainty about parameters in a model.

A number of discrete distributions are related to the *Bernoulli distribution.* A Bernoulli process is basically a trial that has two possible outcomes: a "success" and a "failure." The probability of success, "p," is the only parameter of the Bernoulli distribution. The easiest example to imagine is tossing a coin. But any event where there are two possibilities can be modeled as a

* Other references (such as Bean, M. 2001. *Probability: The Science of Uncertainty.* Belmont, CA: Wadsworth Group) provide detailed analysis and application of a number of these distributions.

Bernoulli distribution—whether or not the government will enact a particular regulation, whether candidate A or candidate B will get elected, whether a drug gets approved by the FDA, whether the inflation rate at the end of the year will be greater than 3% or not, etc.*

When a Bernoulli process is repeated a number of times, or *trials* (n), the resulting distribution for the number of successes out of n trials follows a *Binomial* distribution. For example, if you wanted to model the number of people who drive red cars by calling 100 random phone numbers, the Binomial would be an appropriate choice.

A related distribution is the *Negative Binomial* distribution (also called the Polya distribution in the insurance literature when n is an integer), which gives the number of failures that occur before the rth success is obtained. A potential application would be to model the number of sales calls necessary before five sales are transacted.

The *Geometric* distribution provides the number of failures that occur before the first success is obtained. The *Hypergeometric* distribution provides the number of failures (or defects) obtained from a subset out of a population of D defects (and N – D nondefective items) if items are drawn at random without replacement. All of the preceding distributions are related to the Bernoulli and Binomial processes.

The *Poisson* distribution is conceptually related to the Binomial distribution. While the Binomial distribution assumes that there are n trials, each with a probability of success p, within the Poisson process there is one parameter, lambda (λ), which is the constant rate at which successes occur. The Poisson is used when there is a continuing "exposure" that an event may happen, without the distinct trials of the Binomial distribution. Example situations where a Poisson distribution is often used include modeling the number of accidents that happen within a certain period of time or the number of outbreaks of a disease in a country over the next 5 years.

The *Discrete* distribution is quite different. It can be used to model a number of discrete values, each with some probability of occurring. For example, if future tax rates can take on one of five potential values (due to proposed

* This last example suggests an interesting place to obtain estimates of probabilities for Bernoulli distributions. Online prediction markets, such as Intrade (www.intrade.com), readily provide market-based estimates for probabilities of a large number of financial, political, and social events. Intrade is the largest, but there are a growing number of similar efforts, including a number of companies that use internal prediction markets to estimate probabilities based on the collective judgment of their various stakeholders. There is even a journal, *The Journal of Prediction Markets* (published by the University of Buckingham Press), devoted to the study of such markets. Prediction markets also provide data for distributions other than the Bernoulli, depending on the particular event being marketed (e.g., there may be a series of markets for various ranges for future oil prices that, taken together, can approximate a continuous probability distribution). In addition, there is a long history of applications of the Delphi technique to elicit expert opinion from a group, although Delphi techniques generally attempt to build a consensus opinion out of a diverse group of opinions.

legislation), then the discrete distribution can be used to simulate these outcomes. When the outcomes are sequential integers (e.g., 2, 3, 4, 5) and all have the same probability of occurring, then the *Discrete Uniform* distribution may be used.

The relationship among several of these discrete distributions and their relationship to the Bernoulli process are illustrated in the spreadsheet Discrete4.xlsx. Imagine that we have 1,000 manufactured items, where there is a constant defect probability. (The spreadsheet uses p = .05.) We can simulate each of the 1,000 items, record whether or not it is defective, and calculate the total number of defects, the first defective item, and the number of items without defects before the first occurs. We also calculate the number of defective items out of the first 30 items.

The number of defects is a Binomial (1000, 0.05). The total number of defects can be approximated by using a Poisson distribution (because n*p is large enough), the item with the first defect by using a Negative Binomial (or approximated by an Exponential distribution), the number of defects out of the first 30 items by using a Binomial (30, 0.05) (or approximated by a Hypergeometric distribution), and the number of items before the first defect by using a Geometric distribution. Figure 4.19 shows the portion of the spreadsheet designed to do these calculations and simulations.

The simulation results shown in Figure 4.20 show that these distributions indeed closely resemble the actual calculations from the model.* The shaded cells in the figure provide the means and standard deviations for the calculated cells and the simulated distributions and are the cells to compare.†

4.6 Commonly Used Continuous Distributions

The *Normal* distribution (often referred to as a "bell-shaped curve" or *Gaussian* distribution) is the most widely known distribution. Quite a few naturally occurring phenomena follow a normal distribution. (But, be careful because most do not!) The Central Limit Theorem tells us that the sample mean for a random sample will be normally distributed (provided the sample size

* If you mark the simulated cells as outputs and overlay the distributions, the similarities are even more striking. This also illustrates how simulation may be used to derive probabilistic relationships. The relationships among these distributions can be derived mathematically, but they can also be verified empirically through simulation.

† The unshaded cells should not be compared since they simply show a single simulation from each of these cells. The shaded cells give the means and standard deviations for 10,000 simulations.

	Calculated					Simulated			
		mean	std dev			mean	std dev		
Total #	=SUM(B2:B1001)	=VoseSimMean(F3)	=VoseSimStdev(F3)	Poisson	=VosePoisson(A2*C1001)	=VoseSimMean(J3)	=VoseSimStdev(J3)		
First	=VLOOKUP(1,B2:C21001,2,FALSE)	=VoseSimMean(F4)	=VoseSimStdev(F4)	NegBin	=VoseNegBin(1,A2)+1	=VoseSimMean(J4)	=VoseSimStdev(J4)	Exponential	=VoseExpon(1/A2)
Number Failures	=F4-1	=VoseSimMean(F5)	=VoseSimStdev(F5)	Geometric	=VoseGeometric(A2)	=VoseSimMean(J5)	=VoseSimStdev(J5)		
Failures out 30	=SUM(B2:B31)	=VoseSimMean(F6)	=VoseSimStdev(F6)	Hypergeometric	=VoseHypergeo(30,50,100	=VoseSimMean(J6)	=VoseSimStdev(J6)		

FIGURE 4.19
Several related discrete distributions.

	Calculated					Simulated					
		mean	std dev			mean	std dev				
Total #	63	50.0024	6.903663	Poisson	38	50.0835	7.114562				
First	29	19.9121	19.78814	NegBin	4	19.801	19.25076	Exponential	4.509138692	20.14086	20.2674
Number Failures	28	18.9121	19.78814	Geometric	11	18.9703	19.66577				
Failures out 30	1	1.5034	1.182678	Hypergeometric	3	1.5093	1.17909				

FIGURE 4.20
Discrete distribution results.

is large enough), and this is the basis for much of statistical inference. The Normal distribution has two parameters (the mean and standard deviation) and has the following characteristics: It is symmetric, with about 95% of the probability within two standard deviations of the mean.

The *Lognormal* distribution is also characterized by a mean and a standard deviation, but has a right-hand skew. For example, the Lognormal may be a good choice for modeling the distribution of incomes within a population or the price of a commodity 3 years from now.

The *Student* distribution is similar to the Normal, but with fatter tails. ModelRisk has variants of the Student that permit it to have even fatter tails.

The *Beta* distribution ranges from zero to one, has two parameters, and can take a number of shapes depending on these parameters. It is commonly used to model uncertain proportions or prevalence, such as the fraction of defective items in a sample of manufactured items or uncertainty about the prevalence of a disease.

The *Exponential* distribution has one parameter, β, which is the reciprocal of the mean of the distribution (i.e., = 1/mean). It is also related to the Poisson: It gives the length of time until the first occurrence of an event. For example, the Exponential could be used to model the length of time, *T*, until the first call is made to a call center, while the Poisson could be used to model the probability that no calls would be made during the time interval (0,T) or to forecast how many calls will be made per hour or per day.

The *Gamma* distribution is a generalization of the Exponential and provides the waiting time until the *r*th arrival, when the number of arrivals in each time period follows a Poisson distribution.

The *Pareto* distribution has two parameters and has a similar shape to the Exponential (but with a heavier tail—higher probability of extreme outcomes). The Pareto distribution has the fattest tail of all distributions and is therefore often used in the insurance industry to model extreme events.

Two distributions are commonly used to model the time until the next event—for example, to estimate the uncertainty of lifetimes of gadgets. The *Weibull* distribution also has two parameters (α and β) and provides the lifetime for an item with an instantaneous risk of failure that follows a power function. (When β > 1, this means that the risk of failure will decrease over time.) When β = 1, the Weibull is the same as the Exponential, and the risk of failure is assumed to be constant.

The *Uniform* distribution, which has a minimum and maximum as its two parameters, may be used to model parameters in a model where there is no information to suggest whether some values are more likely than others for a parameter.

The *Triangle* distribution is commonly used to model expert opinion and uses three parameters: the minimum, most likely, and maximum values. We prefer the use of the *PERT* distribution, however, because the mean and standard deviation of the PERT are less sensitive to extreme values for the minimum and maximum. For example, experts may provide extremely small minima or large maxima and the mean of the Triangle distribution is greatly affected by these. In contrast, the PERT distribution is less sensitive to these extremes.

A number of distributions are available for fitting distributions empirically to data without restricting them to any particular functional form (called *nonparametric* distributions). Among these, three commonly used are the *Cumulative* distribution (VoseCumulA or VoseCumulD, depending on whether the data are ascending or descending in order); the *Ogive* distribution (VoseOgive); and the *Relative* distribution (VoseRelative). The Ogive and Relative distributions take as inputs a series of data values and probabilities associated with these values. The Cumulative distributions use as inputs a series of data values and the cumulative probability of observing values less than or equal to these (with the data values either in ascending or descending order). All of these distributions fit the data empirically, without imposing any assumptions regarding the underlying form of the distribution function.

4.7 A Decision Guide for Selecting Distributions

A fundamental choice concerns whether distributions are to be chosen based upon the data or based upon judgment. Even if you have historical data, the choice must be faced. Fitting past data, however perfect and however much uncertainty about the fit is included, can only model what the past data tell us. If there are reasons to expect the future to differ from the past, then some degree of judgment may be called for.* When judgment is used, it may be based on theoretical considerations (such as using an Exponential or Weibull distribution to model waiting times), as in Sections 4.5 and 4.6, or expert opinion, as discussed in Section 4.3.

When fitting data, there are two primary approaches: functional or empirical. Functional fitting entails examining which probability distribution best describes the uncertain process being modeled. Absent data, there is no

* If there are sufficient data, a hybrid approach can be used where a distribution is fit to the data first and then adjustments, based on judgment, are employed to modify the fitted distribution.

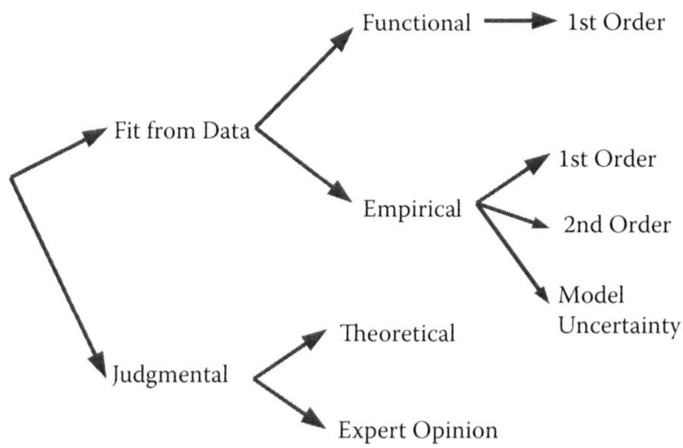

FIGURE 4.21
Decision tree for selecting distributions.

opportunity to incorporate Second Order fitting or model uncertainty. When fitting data empirically, the task is to examine which probability distributions from a set of distribution functions best fit the data. The approach can be First Order, Second Order to represent the lack of precision in the fit, or Bayesian Model Averaging to include uncertainty concerning which distribution best fits the data. Figure 4.21 provides a schematic of the decision process for selecting distributions.

Regardless of the approach chosen, you should visually examine the chosen distribution to ensure that it represents either data or judgment accurately. Sensitivity analysis for the chosen distributions and their parameters is also appropriate in order to understand the extent to which the choice of the distribution matters. As former U.S. Secretary of Defense Donald Rumsfeld stated, "There are known knowns. These are things that we know. There are known unknowns. That is to say, there are things that we now know we don't know. But there are also unknown unknowns. These are things we do not know we don't know."* We can model the known knowns and known unknowns, but modeling the unknown unknowns is a lot harder.

4.8 Bayesian Estimation†

Bayesian approaches and methods to support decision making are increasingly being used. The basic idea is that we have current knowledge (or lack

* February 12, 2002. Department of Defense news briefing, as reported by *Slate* in The poetry of D. H. Rumsfeld, by Hart Seely, April 2, 2003, accessed at www.slate.com on August 28, 2010.

† This section is more advanced and can be skipped without affecting the flow of the rest of the book. However, we think readers will find it a worthwhile effort.

thereof) and, when additional data are collected, we can use these data to update our knowledge. A simple example will illustrate the basic thinking behind this approach and how it can be implemented in a simulation model.

Consider the class of SSRI inhibitors—a common form of antidepression medication. Concerns have been voiced about their use, particularly in children.* Imagine a researcher trying to estimate the rate of an SSRI side effect in children. Such a study is costly; therefore, the researcher starts with a small study in which 50 children are observed on the drug and the rate of suicidal thoughts/actions is recorded. We assume that these side effects occur for 1 out of the 50 children (2%). What does this tell us? It certainly does not tell us that the exact rate of side effects is 2% because we have only studied a few patients. There is uncertainty about what the true rate of side effects is. Statistical theory provides a number of approaches for estimating the true rate, and one of the more flexible is to use a Bayesian approach.

A Bayesian analysis is based on the principle that we learn. The method utilizes Bayes's rule, which shows how our knowledge (or lack of knowledge) is updated as information is gained. In summary, Bayes's rule says that

New Knowledge = Prior Knowledge * Likelihood

This means that we take the prior knowledge and multiply it by the likelihood that we would get what we observed *conditional on the prior knowledge* that we had (called the "likelihood"). The result is the updated knowledge (called the "posterior" probability). We can implement this in a spreadsheet (Bayes4.xlsx), making use of several ModelRisk distributions. Figure 4.22 shows a section of the spreadsheet model where we construct the Bayesian estimates. Since we are trying to estimate a probability, within the model we are evaluating all values between zero and one in the spreadsheet.

Column L lists the potential values for the side effect probability that we will consider. (We have chosen increments of .001 between zero and one. Figure 4.22 only shows a portion of these values; the rows extend to row 1005 to encompass values from zero to one.) Column M is our initial (prior) knowledge; in this case, we have chosen equal probabilities for each value, which is also called a "flat prior." This is commonly assumed if there is no further information upon which to base initial estimates. The likelihood function is column N: It gives the likelihood of getting the number of side effects that we got in our small sample, conditional on each potential side effect probability.† For the posterior probabilities in column O, apply Bayes's rule

* It should be emphasized that, while numerous serious side effects have been documented, most analyses show that the benefits of SSRI use outweigh its costs. In the present example, we only explore one potential negative side effect and this should not be taken as a complete view of the relevant costs and benefits of SSRIs.

† The likelihood function is often complex. In this case, it was the Binomial because each child has a constant probability of having a side effect (i.e., the underlying process is a Binomial process).

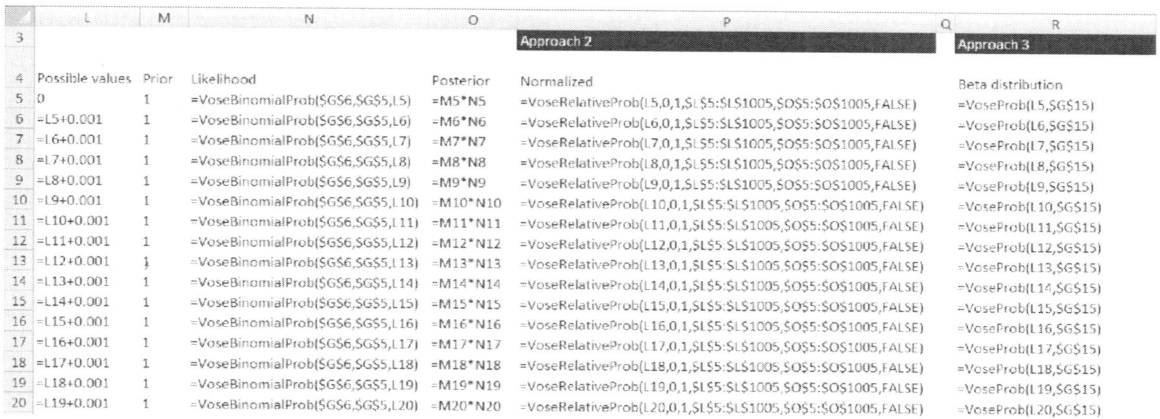

	L	M	N	O	P	Q	R
3					Approach 2		Approach 3
4	Possible values	Prior	Likelihood	Posterior	Normalized		Beta distribution
5	0	1	=VoseBinomialProb($G$6,$G$5,L5)	=M5*N5	=VoseRelativeProb(L5,0,1,$L$5:$L$1005,$O$5:$O$1005,FALSE)		=VoseProb(L5,$G$15)
6	=L5+0.001	1	=VoseBinomialProb($G$6,$G$5,L6)	=M6*N6	=VoseRelativeProb(L6,0,1,$L$5:$L$1005,$O$5:$O$1005,FALSE)		=VoseProb(L6,$G$15)
7	=L6+0.001	1	=VoseBinomialProb($G$6,$G$5,L7)	=M7*N7	=VoseRelativeProb(L7,0,1,$L$5:$L$1005,$O$5:$O$1005,FALSE)		=VoseProb(L7,$G$15)
8	=L7+0.001	1	=VoseBinomialProb($G$6,$G$5,L8)	=M8*N8	=VoseRelativeProb(L8,0,1,$L$5:$L$1005,$O$5:$O$1005,FALSE)		=VoseProb(L8,$G$15)
9	=L8+0.001	1	=VoseBinomialProb($G$6,$G$5,L9)	=M9*N9	=VoseRelativeProb(L9,0,1,$L$5:$L$1005,$O$5:$O$1005,FALSE)		=VoseProb(L9,$G$15)
10	=L9+0.001	1	=VoseBinomialProb($G$6,$G$5,L10)	=M10*N10	=VoseRelativeProb(L10,0,1,$L$5:$L$1005,$O$5:$O$1005,FALSE)		=VoseProb(L10,$G$15)
11	=L10+0.001	1	=VoseBinomialProb($G$6,$G$5,L11)	=M11*N11	=VoseRelativeProb(L11,0,1,$L$5:$L$1005,$O$5:$O$1005,FALSE)		=VoseProb(L11,$G$15)
12	=L11+0.001	1	=VoseBinomialProb($G$6,$G$5,L12)	=M12*N12	=VoseRelativeProb(L12,0,1,$L$5:$L$1005,$O$5:$O$1005,FALSE)		=VoseProb(L12,$G$15)
13	=L12+0.001	1	=VoseBinomialProb($G$6,$G$5,L13)	=M13*N13	=VoseRelativeProb(L13,0,1,$L$5:$L$1005,$O$5:$O$1005,FALSE)		=VoseProb(L13,$G$15)
14	=L13+0.001	1	=VoseBinomialProb($G$6,$G$5,L14)	=M14*N14	=VoseRelativeProb(L14,0,1,$L$5:$L$1005,$O$5:$O$1005,FALSE)		=VoseProb(L14,$G$15)
15	=L14+0.001	1	=VoseBinomialProb($G$6,$G$5,L15)	=M15*N15	=VoseRelativeProb(L15,0,1,$L$5:$L$1005,$O$5:$O$1005,FALSE)		=VoseProb(L15,$G$15)
16	=L15+0.001	1	=VoseBinomialProb($G$6,$G$5,L16)	=M16*N16	=VoseRelativeProb(L16,0,1,$L$5:$L$1005,$O$5:$O$1005,FALSE)		=VoseProb(L16,$G$15)
17	=L16+0.001	1	=VoseBinomialProb($G$6,$G$5,L17)	=M17*N17	=VoseRelativeProb(L17,0,1,$L$5:$L$1005,$O$5:$O$1005,FALSE)		=VoseProb(L17,$G$15)
18	=L17+0.001	1	=VoseBinomialProb($G$6,$G$5,L18)	=M18*N18	=VoseRelativeProb(L18,0,1,$L$5:$L$1005,$O$5:$O$1005,FALSE)		=VoseProb(L18,$G$15)
19	=L18+0.001	1	=VoseBinomialProb($G$6,$G$5,L19)	=M19*N19	=VoseRelativeProb(L19,0,1,$L$5:$L$1005,$O$5:$O$1005,FALSE)		=VoseProb(L19,$G$15)
20	=L19+0.001	1	=VoseBinomialProb($G$6,$G$5,L20)	=M20*N20	=VoseRelativeProb(L20,0,1,$L$5:$L$1005,$O$5:$O$1005,FALSE)		=VoseProb(L20,$G$15)

FIGURE 4.22
Bayesian learning spreadsheet.

	L	M	N	O	P	Q	R
3					Approach 2		Approach 3
4	Possible values	Prior	Likelihood	Posterior	Normalized		Beta distribution
5	0.000	1.000	0.000	0.000	0.000		0.000
6	0.001	1.000	0.048	0.048	2.429		2.428
7	0.002	1.000	0.091	0.091	4.624		4.623
8	0.003	1.000	0.129	0.129	6.604		6.603
9	0.004	1.000	0.164	0.164	8.383		8.381
10	0.005	1.000	0.196	0.196	9.975		9.973
11	0.006	1.000	0.223	0.223	11.395		11.393
12	0.007	1.000	0.248	0.248	12.654		12.652
13	0.008	1.000	0.270	0.270	13.766		13.763
14	0.009	1.000	0.289	0.289	14.740		14.736
15	0.010	1.000	0.306	0.306	15.587		15.583
16	0.011	1.000	0.320	0.320	16.317		16.314
17	0.012	1.000	0.332	0.332	16.940		16.936
18	0.013	1.000	0.342	0.342	17.463		17.459
19	0.014	1.000	0.351	0.351	17.895		17.891
20	0.015	1.000	0.358	0.358	18.243		18.239

FIGURE 4.23
Comparison of the construction method and the Beta distribution.

by multiplying the prior probabilities by the likelihood function. Column P normalizes these probabilities so that the total probability of getting a value between zero and one (i.e., the area below the curve, since this is a continuous distribution) is one. This posterior distribution could become the prior probability if we were to take a second sample.*

The previous paragraph explains how to construct within Excel a posterior distribution, but in this special case there is an easier alternative. Because we are estimating the uncertainty of a proportion based on historical data (in this case, 1 out of 50), we can also use a Beta distribution to get the same results. There are two main input parameters for a Beta distribution: alpha and beta; in this case, alpha would be 1 + 1 = 2 (alpha = number of successes + 1) and beta would be 50 – 1 + 1 = 50 (beta = number of failures + 1).

The simpler approach, using the Beta distribution, is in column R of Figure 4.22. Figure 4.23 shows the same portion of the spreadsheet, with the values in the cells rather than the formulas. As you can see, this figure shows that the construction method gives the same normalized posterior probabilities as the beta distribution; the latter is much easier to implement.

After the initial small study, the researcher is planning a second larger study with 1,000 patients. How many side effects will she or he find? Using the Binomial distribution with $n = 1{,}000$ patients, there are two possible approaches:

* Taking multiple samples proceeds the same way. Each sample's normalized posteriors become the priors for the following sample. This is how continued learning is built into the Bayesian estimates.

- Approach 1. Using p = 1/50 = 2% and thus ignoring the uncertainty we have regarding the results of the first study
- Approach 2. Using the Beta distribution to represent the uncertainty around p

Figure 4.24 shows how to construct both alternative models (with two versions of the second method: the construction approach, Approach 2, and the simpler Beta distribution, Approach 3) to simulate the number of side effects in the second study.

Approach 1 simply uses the Binomial distribution and assumes that the observed side effect rate (.02) from the small study is the actual side effect rate, known with certainty. Approach 2 utilizes Columns L through P, where the priors are updated to derive posterior probability estimates for the potential side effect rate. This (cell G29) relies on using the binomial distribution and the posterior probabilities (represented by the VoseRelative distribution). Approach 3, which should be equivalent to Approach 2, uses the Beta distribution (cell G14) to represent the probability of side effects (in cell G31).

The number of side effects estimated from running the simulation is shown in Figure 4.25. Three things stand out in the figure. First, it appears that the Bayesian approach results in a larger expected number of side effects in the larger study. However, this is somewhat artificial since it really results from our use of the flat prior. Essentially, we are assuming that all side effect probabilities from zero to one are equally likely, so more weight is being given to side effect probabilities larger than .02 than to side effect probabilities smaller than .02.*

Second, the results for Approaches 2 and 3 are almost identical. Third, and most importantly, including the uncertainty about the true probability of a side effect will cause us to have much more uncertainty about the number of side effects in the second follow-up study.† In other words, ignoring this uncertainty can significantly underestimate risk.

Bayesian techniques are increasingly used in diverse fields such as finance, health, pharmaceutical research, and marketing. This chapter provided a brief example of how Bayesian techniques can be (and are) often included in Monte Carlo simulation models.

* If we assume prior probabilities centered around .02, then the difference between the means of the approaches will disappear. However, this would require adapting the Beta distribution (Approach 3) so that it still matches the results from the construction method (Approach 2).

† The boxes in Figure 4.24 represent the middle 50% of the simulations. The width of the box for Approach 1 is approximately 6, while it is 33 for the Bayesian method.

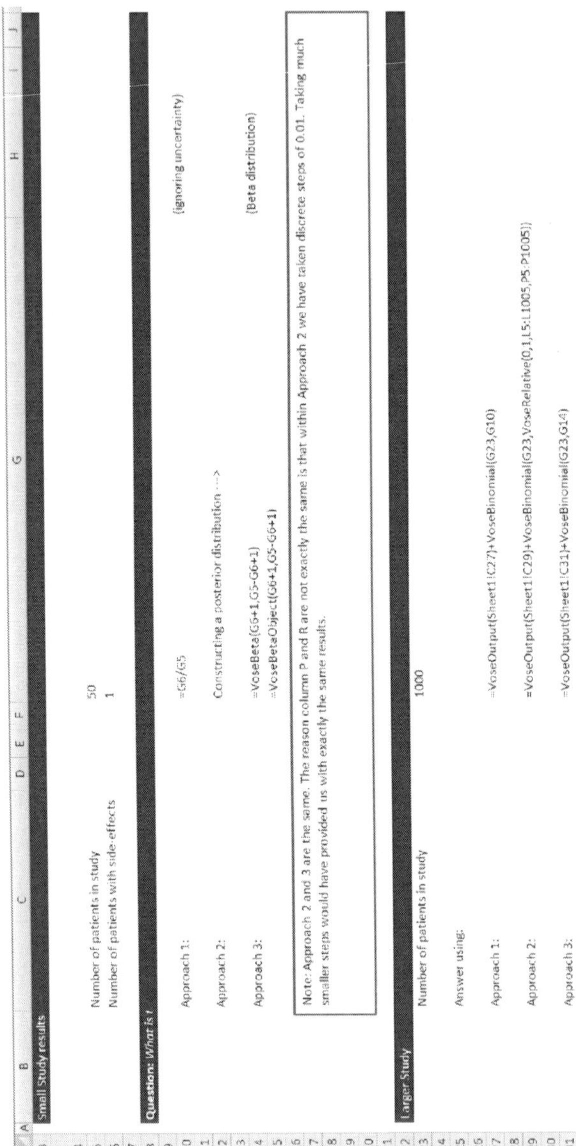

FIGURE 4.24
Non-Bayesian and Bayesian models.

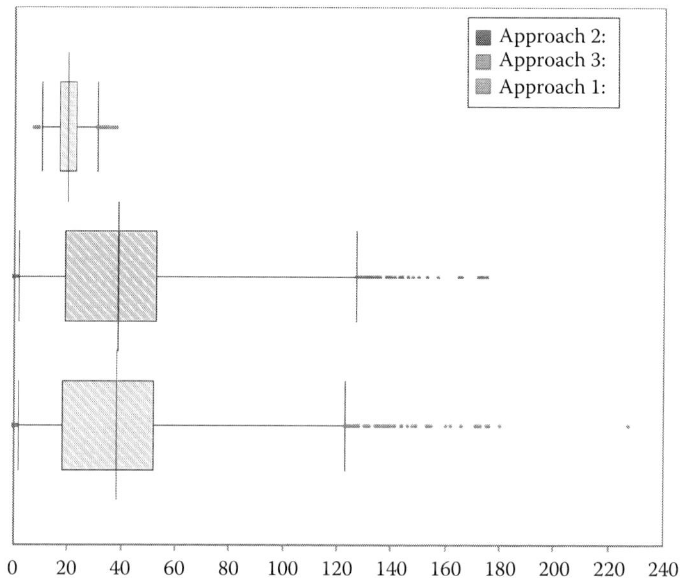

FIGURE 4.25
Estimated prevalence of side effects for Bayesian and non-Bayesian models.

CHAPTER 4
Exercises

4.1 SMOKING AND BIRTH WEIGHT

Exercise4-1.xlsx contains data on the birth weights (in ounces) of 742 babies born to mothers who smoked and 484 babies born to mothers who did not smoke.

a. Fit distributions, with and without uncertainty, to each set of baby birth weights.
b. Low birth weight babies (defined to be less than 88 ounces and greater than or equal to 53 ounces) are at higher risk for a number of health problems. Very low birth weight babies (defined to be less than 53 ounces) are generally premature and at significant risk not to survive. Estimate the risk (with and without uncertainty) of low and very low birth weight babies for smoking and nonsmoking mothers. The *odds* of an event are defined as the ratio of the probability of the event occurring to its not occurring. The *odds ratio* between two groups, then, is the ratio of the odds for the two groups. Express your answer in terms of the odds ratio of having a (very) low birth weight baby for smokers versus nonsmokers.

4.2 AIRLINE DELAYS

The Bureau of Transportation Statistics* provides on-time data for airlines in the United States. Exercise4-2.xlsx contains all flight data for January 2010, and Exercise4-2short.xlsx contains the data for flights departing from Denver International Airport during that month. (Use the larger file if you want to test your database skills for extracting the Denver data; the smaller file already contains the extracted data for those that do not want to download the large file.)

a. Count the number of flights with departure delays (including the cancelled flights) and count the number of flights with arrival delays. Based on the count, determine a probability that a flight will have a departure delay and the probability that a flight will have an arrival delay. Assuming that, during a typical January, there are 20,000 departure flights, build a binomial distribution object to simulate the number of delays.
b. Fit distributions (without parameter uncertainty) for the length of the departure delays and the length of the arrival delays (contingent on there being a delay). Choose distributions that visually appear to fit the data best.
c. Estimate the total monthly departure and arrival delays in minutes of flights leaving DIA (for a "typical" January). Provide a 90% confidence interval for this estimate.

4.3 INVESTMENT RETURNS

Exercise4-3.xlsx provides the annual returns to stocks (measured by the S&P 500), Treasury bonds, and Treasury bills over the 1929–2009 period. Consider an investment of $100 invested in each of these assets in 2010.

a. Fit a distribution (without parameter uncertainty) to each of these assets' annual returns. Choose distributions based on the best AIC score.
b. Simulate the value of a $100 investment in each asset by year over a 10-year period.
c. What is the probability of losing money in the investment in each asset over each time period?
d. What is the first percentile in each case?
e. Repeat parts (a) through (d), but select distributions that visually appear to fit the data best.
f. Repeat parts (a) through (d), using the empirical Ogive distribution.
g. Compare the three approaches. Try using overlay charts in the comparison. Why are AIC best fits so different from the empirical fits?
h. Optional: Repeat all of the preceding parts, using parameter uncertainty in the fitted distributions.

---

* www.transtats.bts.gov.

4.4 REAL ESTATE CRISIS

In October 2010, the collapse of the U.S. real estate market was continuing. Exercise4-4.xlsx contains data for single-family homes from major cities showing the fraction of homes that had cut their price, the median listing price, and the year-to-year change of listed prices.* You are interested in assessing the probability that home prices are still decreasing, based solely on these data.

a. Consider two measures of how prices are changing: the fraction of homes that have cut their listing price and the last year's percent change in the median listed price. Fit distributions (based on the best HQIC score) to each of these measures and provide an assessment of the average, median, and fifth percentile of these distributions. Include parameter uncertainty.
b. Repeat part (a), but choose distributions based on the best SIC score.
c. Use Bayesian Model Averaging to include model uncertainty for the fitted distributions in parts (a) and (b).

4.5 BROADBAND INTERNET USAGE

Internet usage continues to grow rapidly. Most residential consumers in developed countries now access the Internet through a broadband connection. Recent data reveal that usage patterns vary markedly across residential subscribers: In 2009, the top 1% of users accounted for 25% of total traffic. Average usage was 9.2 GB/month, while the median usage was 1.7 GB/month. In addition, growth rates per user were 30%–35%. Further detail on the distribution of usage across subscribers is provided in Table 4.1.†

a. Using these data, create a table showing the cumulative distribution for usage. Then simulate usage levels by creating a Step Uniform distribution (for percentiles ranging from 0% to 100% in 1% steps) and using the output of the Step Uniform, along with a VLOOKUP function and your cumulative distribution table, to obtain a distribution of individual usage levels. Also, fit a Lognormal distribution using the mean and median usage levels stated previously. (Do not use the table for this; you will have to try various standard deviations manually until the median matches the given information.) Run a simulation of each distribution and overlay the results. (The cumulative distribution provides the best view of this.)

---

* The data are for August 2010 and come from www.zillow.com, accessed on October 18, 2010.

† *Broadband performance.* OBI technical paper no. 4, Federal Communications Commission, 2010. We have adapted the numbers slightly in order to merge data from different parts of the report and to make it more suitable for analysis (for example, by eliminating ranges of usage levels—especially the last usage category, which had been open ended at 15+ GB).

TABLE 4.1

Broadband Internet Usage Patterns

Monthly Usage Level (GB)	Percentage of Users
0.1	12
0.3	18
0.85	17
1.65	18
3.2	13
5.1	10
25	8
75	3
250	1

b. One disadvantage of tables like this is that only discrete steps of information are given. For example, there have been proposals to cap individual Internet usage at 5 GB per month; usage levels above this would entail surcharges. Table 4.1 does not readily show what percentage of users would be impacted by such a cap. Use your Lognormal distribution from part (a) to estimate this percentage.

c. Assume that annual growth in each of the next 3 years will follow a Uniform distribution with a minimum of 30% and a maximum of 35%. Simulate 3 years of growth in broadband usage levels. It is not obvious how to do this with the Lognormal distribution you estimated. (It is easy to have the mean grow, but what do you do with the standard deviation?) Use Table 4.1 and grow each usage level for 3 years at a rate given by the Uniform distribution (30%, 35%) and then simulate the usage level from this new table. Note that you cannot assume that the growth will be exactly the same every year. Based on that simulation, use the mean and standard deviation in a Lognormal distribution to simulate individual usage patterns in 3 years' time. What percentage of users would be subject to the 5 GB cap at that time? What percentage would be subject to the cap if it were to grow at the same rate each year as the usage levels?

4.6 ALTERNATIVE ENERGY

There are a number of Internet-based expert opinion surveys. One, using 97 experts, collected predictions that alternatives to carbon-based fuel will provide 30% of all energy used worldwide.* The 97 experts predict an average year of 2022 for this to happen, with a standard deviation of 6 years. More detailed data are provided in Table 4.2.

* www.techcast.org/BreakthroughAnalysis.aspx?ID=27, accessed on October 17, 2010. The data in the question have been extracted and modified somewhat for this exercise.

TABLE 4.2

Expert Opinions on When Alternative Energy Equals 30% of the Total Energy Use

Most Likely Year	Number of Experts
2010–2013	2
2014–2017	9
2018–2021	41
2022–2025	22
2026–2029	3
2030–2033	13
2034–2038	7

a. Choose an appropriate probability distribution to reflect the most likely year. Use this distribution to simulate the year.
b. Provide your own probability distribution for the most likely year. Combine your distribution with that of the experts, weighting the expert opinion by 70% and your own by 30%. Simulate again and provide a predicted mean for the most likely year.

4.7 WARRANTIES

Tires can fail for a number of reasons; consider three: tread wearout, sidewall failure, or steel shear of tread. Sidewall failures and steel shear of tread follow exponential distributions, with means equal to a design mileage chosen by the manufacturer. Tread wearout follows a normal distribution, with the same design mean and a standard deviation of 5,000 miles. The tire warranty mileage is set at 10,000 miles less than the design mileage. The warranty will cover all three types of damage, but does not cover punctures; these follow an exponential distribution (with mean time to failure of 35,000 miles). In addition, the warranty is void if the car is sold to another owner; this follows an exponential distribution with a mean of 75,000 miles. Assume that a damaged tire must be replaced and consider only originally purchased tires (not tires replaced after a warranty-covered failure—i.e., repeat failures).

a. The tire manufacturer can design tires for mean failures of 40,000, 60,000, or 110,000 miles. Estimate the probability of at least one warranty-covered failure for each design option (i.e., the probability of a covered failure occurring at a mileage less than the warranty mileage but before the car is sold).
b. Suppose that the cost of producing the tires is $30 each for the 40,000-mile design, $50 for the 60,000-mile design, and $75 for the 110,000-mile design. A warranty replacement costs the manufacturer the cost of the new tire plus $30 for labor. The tires can be sold for $65, $95, and $120, respectively. Which design has the highest mean profit?

4.8 LET'S MAKE A DEAL

Monty Hall was the host of the long-running television game show, "Let's Make a Deal." The Monty Hall problem is a well studied probability problem that continues to confound people with the results. At the end of each show, the day's winner had a final opportunity to win a great prize. He or she had to select one of three doors: Only one of the doors had the great prize behind it; the other two had nothing valuable (often called "goats"). After the contestant chose a door, Monty always revealed one of the two goats (never what was behind the door chosen by the contestant) and offered the contestant the opportunity to change his or her choice. Should the contestant have changed his or her choice?

a. Build a simulation model to illustrate the two possible strategies: sticking with the initial choice or changing the initial choice.
b. Estimate the probability of winning the great prize under each strategy.

5

Modeling Relationships

LEARNING OBJECTIVES

- Appreciate the importance of including and accurately reflecting key relationships in your models.
- Learn how logical functions can be used to capture important dependencies.
- Learn how to include simple correlation structures between variables.
- Learn how to capture more complex correlations through the use of copulas.
- Learn how to fit correlation patterns to data.
- Learn how to use regression models to capture relationships in your models.
- See an example of how to model the uncertainties inherent in regression models.
- See how the Envelope method can be used to combine expert opinion and logical relationships between variables.

5.1 First Example: Drug Development

Pharmaceutical companies invest considerable sums of money and time in the highly risky process of developing new drugs. Drug development goes through a number of stages, both for product development and regulatory approval, with each stage subject to many uncertainties, including whether the drug appears to be successful at treating a condition and whether there are adverse side effects that would prevent ultimate approval for use. These different development stages are clearly related: A company will only undertake the next stage of development if previous stages were all successful.

These types of conditional and logical relationships are perhaps the simplest and most frequently encountered when building models. Most can be modeled using the variety of logical functions available within Excel: principally the IF, OR, AND functions and a number of related derivatives (such as VLOOKUP, COUNTIF, etc.). When we consider our simple

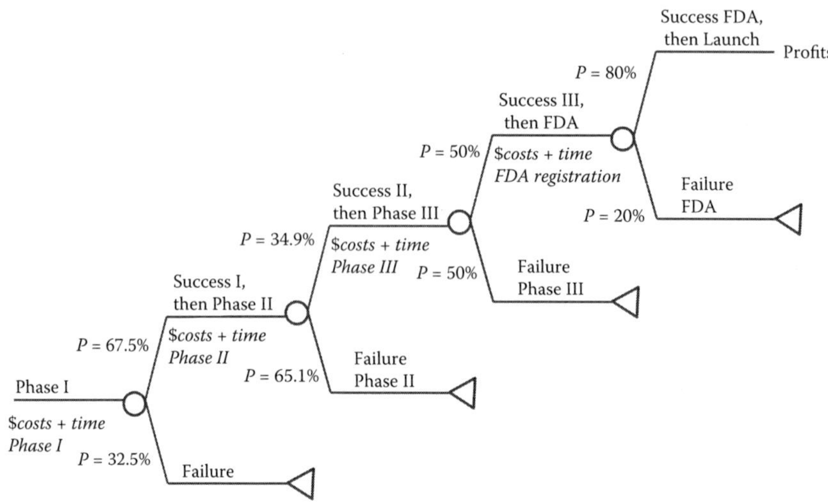

FIGURE 5.1
The drug development process.

drug development example, these types of conditional relationships can be modeled using a combination of Bernoulli distributions and IF statements. Figure 5.1 shows this logic of the drug development process, formatted as a simple decision tree.

Figure 5.1 shows a typical drug going through three stages (phases I to III), followed by registration and potential approval by the FDA. This last step is, of course, conditional on it having survived all three phases. The probabilities are from DiMasi et al.* for the therapeutic class of gastrointestinal/metabolism drugs, based on self-originated compounds first tested on humans from 1993 to 2004.

Each development phase has an associated (uncertain) cost and (uncertain) duration. The costs and durations are modeled using PERT distributions, and success (or failure) at each phase is modeled using a Bernoulli distribution.† Potential profit for the drug is also modeled using a PERT distribution.‡ The model is shown in Figure 5.2 and the essential logic of the model is shown in Figure 5.3.

* DiMasi, J. A., L. Feldman, A. Seckler, and A. Wilson. 2010. Trends in risks associated with new drug development: Success rates for investigational drugs. *Nature* 87:3.

† When n = 1 for the number of trials, the Binomial and Bernoulli distributions are the same thing.

‡ Profits are actually much more complex, depending on patient population, disease prevalence, market competition, costs, etc. The model could be expanded to include these factors, but we just use the PERT distribution so that we can focus on the logical relationships between phases of drug development.

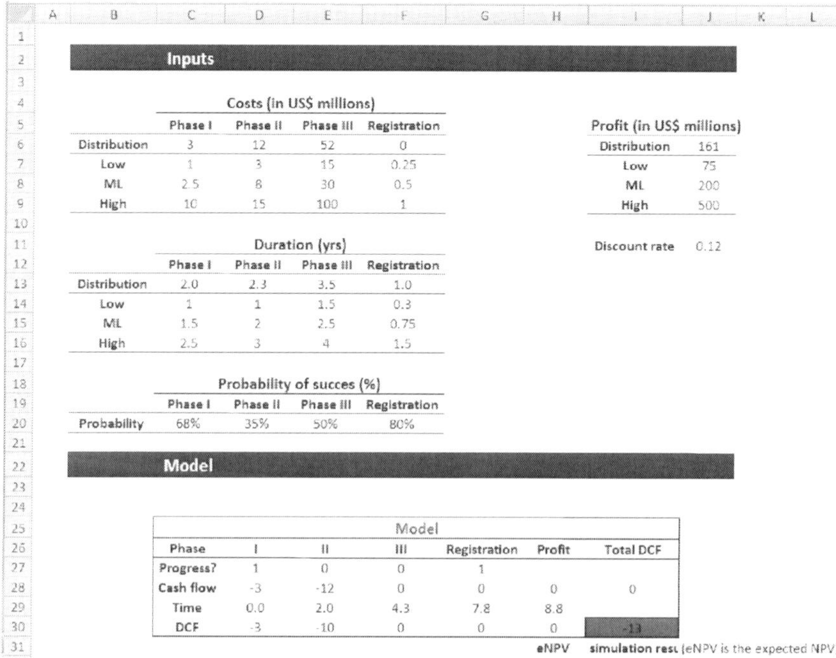

Inputs

Costs (in US$ millions)	Phase I	Phase II	Phase III	Registration
Distribution	3	12	52	0
Low	1	3	15	0.25
ML	2.5	8	30	0.5
High	10	15	100	1

Profit (in US$ millions)	
Distribution	161
Low	75
ML	200
High	500

Discount rate 0.12

Duration (yrs)	Phase I	Phase II	Phase III	Registration
Distribution	2.0	2.3	3.5	1.0
Low	1	1	1.5	0.3
ML	1.5	2	2.5	0.75
High	2.5	3	4	1.5

Probability of succes (%)	Phase I	Phase II	Phase III	Registration
Probability	68%	35%	50%	80%

Model

Phase	I	II	III	Registration	Profit	Total DCF
Progress?	1	0	0	1		
Cash flow	-3	-12	0	0	0	0
Time	0.0	2.0	4.3	7.8	8.8	
DCF	-3	-10	0	0	0	-13
					eNPV	simulation resu (eNPV is the expected NPV)

FIGURE 5.2
Drug development model.

While each phase's success or failure is simulated using the Bernoulli distribution, the formulas in row 28 ensure that there is only a cash flow if the current phase and all previous phases are successful. If any phase's progress produces a zero (no success), all subsequent stages will show that there are no costs, revenues, and cash flow. Similarly, the profit calculation (column H, not shown in Figure 5.3) contains an IF statement that requires D27*E27*F27*G27=1. All phases and registration must be successful in order to launch the drug and thus generate a profit.

The IF statements capture the relationships between the uncertainties about whether the drug will progress through each stage of the development process. Alternatively, we could have used a combination of nested functions within a single cell—IF(AND(D27=1,E27=1,F27=1,G27)—to model the dependence of product launch and profit on the success at all previous development stages. Multiplying the outcomes (either zero or one) of the cells modeling each previous stage accomplishes the same thing, but is less prone to entry errors.

Our model contains 13 probability distributions (four uncertain costs, four durations, four success or failure events, and the uncertain profit—all marked as ModelRisk inputs) and our output is the expected NPV (often

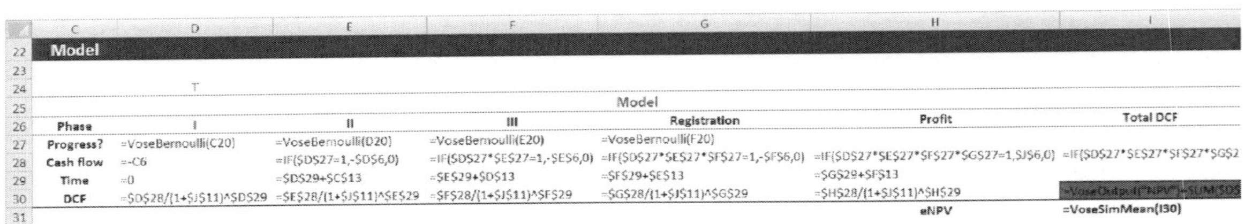

FIGURE 5.3
Use of IF statements in the drug development model.

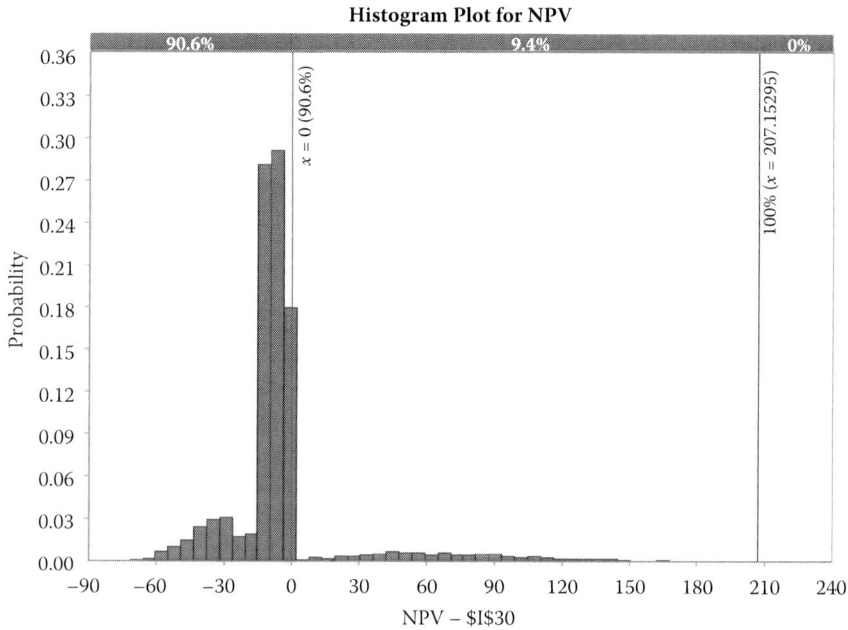

FIGURE 5.4
Results of the drug development model.

called eNPV) of this drug development.* Figure 5.4 provides the resulting simulation results for the NPV: The probability of losing money is over 90%.†

The expected NPV is only −$4.3 million, but the distribution is quite skewed. (The simulated NPV ranges from −$78 million to +$207 million.) Also evident in Figure 5.4 is that the NPV has multiple peaks, which correspond to the various times that the drug can fail during its development. The Tornado chart in Figure 5.5 shows the sensitivity of the expected NPV to the various uncertain factors.

Most critical to the variability of the NPV are phase III and Registration success/failure, closely followed by phase I outcomes, phase I costs, and phase II outcomes. The other inputs have less influence on the NPV, some

* We have assumed a fixed discount rate—often a difficult input to choose for a profitability analysis over time. Some companies adopt a specific "hurdle rate" to discount future costs and revenues, while sometimes an opportunity cost of capital (either an appropriate borrowing or lending rate) approach is used. If it is not clear what the appropriate discount rate is, then we recommend using a simulation table (or other type of "what if" analysis) to analyze how the eNPV is affected by the choice of discount rate. We advise against making the discount rate a probability distribution, however. It is not random in the same sense as the other uncertainties being modeled.

† Notice that the probability of success at all stages = 68%*35%*50%*80% = 9.52%, so the rest of the time (90.48%) drug development will be unprofitable.

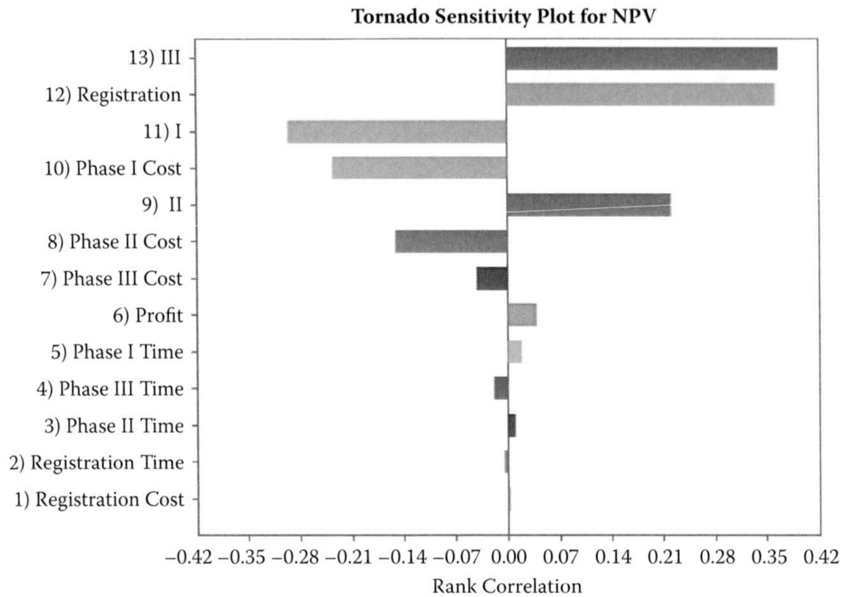

FIGURE 5.5
Tornado chart for drug development.

hardly affecting it at all. This provides guidance about where to concentrate resources to have the greatest impact on the NPV.

As this model demonstrates, the use of logical statements is a simple, powerful, and often preferred way to incorporate dependencies between uncertain factors.

5.2 Second Example: Collateralized Debt Obligations

Collateralized Debt Obligations (CDOs) lie at the heart of the financial meltdown of 2007–2008. CDOs are asset-backed securities that comprised a number of underlying assets* and are divided into different *tranches* having different risk characteristics.† Junior tranches default when one (or only a few) of the underlying assets default. Senior tranches only default when

* In this case, the asset owned by the CDO may be mortgages or other debt instruments that have obligation to make payments to the CDO.

† Each tranche, or slice, of the CDO is made up of a subsection of the underlying assets, and the tranche is then defined based on its level of aggregate risk. Theoretically, a less risky senior tranche may pay less income but is considered to have a lower probability of those assets becoming delinquent or defaulting.

all of the underlying assets default. In this way, by combining a number of different risky assets, the senior tranches may be far less risky than the underlying assets. Individual tranches were then sold as single investments. Numerous CDOs were created, and senior tranches earned top ratings (for low risk) even when the underlying assets were quite risky.* CDOs originated in 1987, but grew rapidly, especially after 2000. Unfortunately, many CDOs were backed by subprime mortgage bonds, and rising defaults in 2007 were a major contributing factor to the global financial crisis. Global CDO issuance reached over $520 billion in 2006 and fell to $6.4 billion in 2010.

A key factor determining the risk of a CDO is the degree of dependence of the underlying assets. Independence would permit a large number of quite risky assets to be combined to produce senior tranches with a very low aggregate risk since default would require simultaneous default of all of the underlying assets. This, of course, requires that the assets truly be independent. Conversely, if the underlying asset risks are highly correlated, then the senior tranche may not be much less risky than the individual assets. During the years before the crisis, it was thought that mortgage risks were geographically uncorrelated: Default rates in one region were largely independent of default rates in other regions. This belief turned out to be incorrect. As one expert observes:

> However, if not used properly, Monte Carlo models can be hazardous to your health. Their results are unreliable if you do not accurately estimate the correlation between variables. These estimates require a sound grasp of underlying causative relationships. To those who miss this point, the models produce an illusion of precision, even though they are built upon the imprecision of random outcomes. Misuse of Monte Carlo models has contributed to some of the biggest Black Swan events of recent years, including the 1998 failure of Long-Term Capital Management (LTCM) and, just a few years later, the collapse of the multi-hundred billion dollar correlation trading market.†

To illustrate this further, we constructed a simplified model of two assets with risk characteristics of junk bond status (i.e., high risk). Two tranches are offered; the junior tranche defaults if either asset defaults, while the senior trance defaults only when both assets default. The spreadsheet CDO5.xlsx contains our model. Annual default probabilities and a ratings schedule are shown in columns G and H. The simulation of individual security default rates and their associated ratings, along with those for the CDOs are shown in Figure 5.6.

---

* In fact, nearly 70% of CDOs were rated AAA while less than 1% of single-name corporate bonds are rated AAA (SIFMA. 2009. IMF Global Financial Stability Report).

† Poser, K. A. 2010. Stalking the Black Swan, 178–179. New York: Columbia Business School Publishing. This book contains a detailed account of the problems that correlation of housing markets created for CDOs.

	A	B	C	D	E	F	G
1	Individual Securities	1	2				annualized default rate
2	Default Rate	=G15	=G15				0
3	Default?	=VoseOutput()+VoseBernoulli(B2,B25)	=VoseOutput()+VoseBernoulli(C2,C25)				0.0002
4	probability	=VoseSimMean(B3)	=VoseSimMean(C3)				0.0006
5	Rating	=VLOOKUP(B4,$G$2:$H$22,2,TRUE)	=VLOOKUP(C4,$G$2:$H$22,2,TRUE)				0.0009
6							0.0012
7	Tranches	Default?	probability	Rating			0.0017
8	Junior	=VoseOutput()+IF(B3+C3>=1,1,0)	=VoseSimMean(B8)	=VLOOKUP(C8,G3:H22,2,TRUE)			0.0019
9	Senior	=VoseOutput()+IF(B3+C3=2,1,0)	=VoseSimMean(B9)	=VLOOKUP(C9,G2:H22,2,TRUE)			0.0025
10							0.0032
11	simulation	Correlation	Junior tranche prob	Junior rating	Senior tranche prob	Senior rating	0.0038
12	1	0	=VoseSimMean($B$27,$A12)	=VLOOKUP(C12,$G$2:$H$22,2,TRUE)	=VoseSimMean(B$28,A12)	=VLOOKUP(E12,$G$2:$H$22,2,TRUE)	0.0075
13	2	0.1	=VoseSimMean($B$27,$A13)	=VLOOKUP(C13,$G$2:$H$22,2,TRUE)	=VoseSimMean(B$28,A13)	=VLOOKUP(E13,$G$2:$H$22,2,TRUE)	0.0107
14	3	0.2	=VoseSimMean($B$27,$A14)	=VLOOKUP(C14,$G$2:$H$22,2,TRUE)	=VoseSimMean(B$28,A14)	=VLOOKUP(E14,$G$2:$H$22,2,TRUE)	0.0145
15	4	0.3	=VoseSimMean($B$27,$A15)	=VLOOKUP(C15,$G$2:$H$22,2,TRUE)	=VoseSimMean(B$28,A15)	=VLOOKUP(E15,$G$2:$H$22,2,TRUE)	0.0204
16	5	0.4	=VoseSimMean($B$27,$A16)	=VLOOKUP(C16,$G$2:$H$22,2,TRUE)	=VoseSimMean(B$28,A16)	=VLOOKUP(E16,$G$2:$H$22,2,TRUE)	0.0259
17	6	0.5	=VoseSimMean($B$27,$A17)	=VLOOKUP(C17,$G$2:$H$22,2,TRUE)	=VoseSimMean(B$28,A17)	=VLOOKUP(E17,$G$2:$H$22,2,TRUE)	0.0324
18	7	0.6	=VoseSimMean($B$27,$A18)	=VLOOKUP(C18,$G$2:$H$22,2,TRUE)	=VoseSimMean(B$28,A18)	=VLOOKUP(E18,$G$2:$H$22,2,TRUE)	0.043
19	8	0.7	=VoseSimMean($B$27,$A19)	=VLOOKUP(C19,$G$2:$H$22,2,TRUE)	=VoseSimMean(B$28,A19)	=VLOOKUP(E19,$G$2:$H$22,2,TRUE)	0.0568
20	9	0.8	=VoseSimMean($B$27,$A20)	=VLOOKUP(C20,$G$2:$H$22,2,TRUE)	=VoseSimMean(B$28,A20)	=VLOOKUP(E20,$G$2:$H$22,2,TRUE)	0.0664
21	10	0.9	=VoseSimMean($B$27,$A21)	=VLOOKUP(C21,$G$2:$H$22,2,TRUE)	=VoseSimMean(B$28,A21)	=VLOOKUP(E21,$G$2:$H$22,2,TRUE)	0.147
22	11	1	=VoseSimMean($B$27,$A22)	=VLOOKUP(C22,$G$2:$H$22,2,TRUE)	=VoseSimMean(B$28,A22)	=VLOOKUP(E22,$G$2:$H$22,2,TRUE)	0.2996
23							
24	This Simulation	=VoseSimTable(B12:B22,,0)					
25	Simulating Copula	=VoseCopulaBiNormal(B24)	=VoseCopulaBiNormal(B24)				
26							
27	Junior Tranche	=VoseOutput()+IF(B3+C3>=1,1,0)					
28	Senior Tranche	=VoseOutput()+IF(B3+C3=2,1,0)					

FIGURE 5.6
CDO default model.

Default for individual securities is modeled using Bernoulli distributions; we have chosen default rates of .0204 for each security, matching a BB– rating (just above junk bond status). After running the simulation, the mean default rate for each security appears in cells B4 and C4 and the associated ratings are provided using VLOOKUP functions. LOOKUP functions are a useful way of relating variables when the value that a variable takes depends on where it lies within a range of possible values. In this case, the VLOOKUP function starts with the mean default rate, finds the lowest default rate in the ratings table that is larger,* and returns the value shown in the second column of the ratings table (the actual rating associated with that default rate).

Modeling the default rates for the CDOs is accomplished using logical IF statements. The Junior tranche will default if either underlying asset defaults, while the Senior tranche only defaults when both assets default.†

All that remains is to examine how correlation between the two underlying assets can affect the resulting CDO default rates. We build a Simtable to permit the simple linear correlation (modeled with a copula, which is explained later in this chapter) between the two assets, as represented by a value called covariance (that ranges from zero to one).‡ We can think of the IF function as modeling a perfect correlation between variables: if A, then B. Correlations that are less than perfect are equivalent to saying something like, "if A, then sometimes B" or "if A is higher, then B tends to be higher or lower." The frequency with which these events occur together reflects the degree to which they are correlated. For positive correlations, the linear correlation can range between zero (independent events) and one (perfectly correlated events that would be better modeled using IF statements).

For each potential correlation, we estimate the probability of default and associated rating for the Junior and Senior tranches in columns C through F next to the Simtable. We estimate this probability by looking at how often it defaults and using the VoseSimMean to estimate the fraction of time that this occurs. Figure 5.7 shows the way that simple linear correlation is represented in ModelRisk. Row 25 contains a VoseCopulaBiNormal function that represents a linear correlation between two variables.§ Referencing the Simtable causes the copula cells (B25 and C25) to vary based on the correlation listed in the Simtable.

The final step is to model the default probabilities of the tranches using the listed correlations between the assets. In ModelRisk, this is accomplished by

* TRUE in the VLOOKUP function finds the closest value that is greater than the value that we are checking; FALSE would look for an exact match.

† Alternatively, we could have used an IF(OR( )) function for the junior tranche default and the IF(AND( )) function for the senior tranche default. Instead, we check whether the sum of defaults for the two assets is greater than 1 or equals 2, respectively, to model the CDO tranche defaults. The results will be identical, but we prefer to avoid combinations of logical statements where possible, due to the ease of making errors when using nested logical functions.

‡ We consider more complex forms of correlation in the next section.

§ We examine copulas more fully in the next section. The use of the Normal copula, here, is equivalent to simple linear correlation.

	A	B	C
24	This Simulation	=VoseSimTable(B12:B22,,0)	
25	Simulating Copula	=VoseCopulaBiNormal(B24)	=VoseCopulaBiNormal(B24)
26			
27	Junior Tranche	=VoseOutput()+IF(B3+C3>=1,1,0)	
28	Senior Tranche	=VoseOutput()+IF(B3+C3=2,1,0)	

FIGURE 5.7
Implementing CDO default correlation with a copula.

adding the second argument in the individual default probability functions that appear in cells B3 and C3 of the spreadsheet. This second argument, referred to as a *U parameter* in ModelRisk, ensures that the simulations in these two cells are correlated. A U parameter ranges between zero and one and represents a percentile of a distribution function.* The two U values are simulated in the two CopulaBiNormal function cells, and degree to which the two U values move "in tandem" is the result of the covariance (i.e., the degree to which the two asset default rates are correlated). By using the copulas as arguments in the Bernoulli functions that simulate the individual asset default rates, these default rates will also reflect the same degree of correlation between them.

Running the simulation provides the results shown in Figure 5.8. Cells B5 and C5 show that the individual securities mean ratings are BB– and BB, just above junk status. The Simtable shows that the junior tranche ratings (in column D) are not much affected by correlation between the assets. However, the senior tranche ratings (column F) are quite impacted by correlation: When there is no correlation between the two underlying assets (shown in row 12), the CDO has an AAA rating, but when the two assets are perfectly correlated (shown in row 22), then the CDO senior tranche is rated BB–. The protection of the senior tranche comes from the requirement that both assets must default before the senior tranche holders are liable; therefore, when the assets are perfectly correlated, there is no extra protection compared to the junior tranche. But when the assets are uncorrelated, then bundling high-risk assets can produce a very high-quality (low-risk) senior tranche CDO. A contributing factor to the financial meltdown of the late 2000s was the belief that the underlying assets in these CDOs were uncorrelated when, in fact, they were correlated.

* When a U parameter is present in a ModelRisk distribution function, the U parameter forces the function to return the percentile given by the U parameter. For example, =VoseNormal(0,1) will generate a random sample from the standard Normal distribution, but =VoseNormal(0,1,.95) will always return 1.645, which is the 95th percentile of the Normal (0,1) distribution. The bivariate copula generates two correlated random variables between zero and one. When those two values are used as the U parameters for two distribution functions in ModelRisk, the two distributions will now generate samples correlated in exactly the same fashion as the copula.

	A	B	C	D	E	F
1	Individual Securities	1	2			
2	Default Rate	0.0204	0.0204			
3	Default?	0	0			
4	probability	0.02	0.02			
5	Rating	BB-	BB			
6						
7	Tranches	Default?	probability	Rating		
8	Junior	0	0.04	B		
9	Senior	0	0.00	AAA		
10						
11	simulation	Correlation	Junior tranche	Junior rating	Senior tranche	Senior rating
12	1	0	0.03830	B	0.00050	AAA
13	2	0.1	0.03820	B	0.00050	AAA
14	3	0.2	0.03860	B	0.00120	AA-
15	4	0.3	0.03860	B	0.00160	AA-
16	5	0.4	0.03700	B	0.00210	A
17	6	0.5	0.03650	B	0.00360	BBB+
18	7	0.6	0.03650	B	0.00470	BBB
19	8	0.7	0.03500	B	0.00610	BBB
20	9	0.8	0.03270	B	0.00830	BBB-
21	10	0.9	0.02950	B+	0.01150	BB+
22	11	1	0.02130	BB-	0.02130	BB-

FIGURE 5.8
CDO default results.

5.3 Multiple Correlations

In order to model more than two assets, using an approach where we had to correlate each asset manually with every other (as with IF statements) would take a lot of time. For example, with six assets, there are 5 + 4 + 3 + 2 + 1 = 15 relationships. In those cases of many-to-many correlations, it is useful to build a *correlation matrix*, which shows the correlations among the various assets. Figure 5.9 shows a possible correlation matrix for six assets.

Each cell of the matrix shows an assumed correlation between a pair of assets. (Blank cells are zero—that is, independent assets.) Notice that this correlation matrix cannot be correct because, while A and C are shown to be independent of each other, the relationships indicated among the six assets mean that A and C must actually be related to some extent through their intermediate correlations with other assets. To capture the relationships among the six assets that are implied by the original correlation matrix, we

	A	B	C	D	E	F	G
1	Defined correlations						
2		A	B	C	D	E	F
3	A	1	0.6		0.8		
4	B	0.6	1			0.9	0.6
5	C			1		0.4	0.5
6	D	0.8			1		
7	E		0.9	0.4		1	
8	F		0.6	0.5			1

FIGURE 5.9
A correlation matrix.

	A	B	C	D	E	F	G
1	Defined correlations						
2		A	B	C	D	E	F
3	A	1	0.6		0.8		
4	B	0.6	1			0.9	0.6
5	C			1		0.4	0.5
6	D	0.8			1		
7	E		0.9	0.4		1	
8	F		0.6	0.5			1
9	Valid Correlation Matrix						
10		A	B	C	D	E	F
11	A	1	0.452884	-0.03955	0.734847	0.069335	0.053538
12	B	0.452884	1	0.060151	0.051447	0.696404	0.462757
13	C	-0.03955	0.060151	1	0.022945	0.330949	0.444655
14	D	0.734847	0.051447	0.022945	1	-0.04023	-0.03106
15	E	0.069335	0.696404	0.330949	-0.04023	1	0.063667
16	F	0.053538	0.462757	0.444655	-0.03106	0.063667	1

FIGURE 5.10
Correcting the correlation matrix.

use the VoseValidCorrmat function to correct the original correlation matrix. This is an Excel array function; to enter it, you highlight the full range of cells in which the function output should appear (B11:G16) and then, while leaving those cells highlighted, click the formula bar and enter the function =VoseValidCorrmat(B3:G8), whose argument (B3:G8) is the original correlation matrix in Figure 5.9, and press Shift + Ctrl + Enter simultaneously. This gives Figure 5.10.

Notice in the new valid/correct correlation matrix that there is some implied correlation between each pair of assets, despite the absence of many pairwise correlations in the original assumed correlation matrix. Now that we have a valid correlation matrix, we can use it in a model by taking advantage of the MultiNormal copula shown in cells I11:I16 of the Correlations5.xlsx spreadsheet. (This is also entered as an array function; alternatively, this copula can be entered using the Correlation → Multivariate copula dialog from the ModelRisk toolbar, which will be covered in the next section.) This Multivariate copula produces six correlated U values, which can then be used to correlate six distributions when simulating the model. For example, in the model we have simulated six asset default rates, each with a .1 probability of default. Column K simulates whether each asset defaults, and cell I19 provides the mean default rate for the first asset. Cell I23 shows the mean rate for simultaneous default of all six assets.

Figure 5.11 shows the completed spreadsheet. The probability for the first asset to default is .1062, but only .0002 for all six to default simultaneously. This should be compared with the probability that all six would simultaneously default if they were independent: $(.1)^6 = .000001$. As you can see, in situations when two or more variables are correlated, including correlations in your model matters.

	A	B	C	D	E	F	G	H	I	J	K	L
1	Defined correlations											
2		A	B	C	D	E	F					
3	A	1	0.6		0.8							
4	B	0.6	1			0.9	0.6					
5	C			1		0.4	0.5					
6	D	0.8			1							
7	E		0.9	0.4		1						
8	F		0.6	0.5			1					
9	Valid Correlation Matrix											
10		A	B	C	D	E	F		The copula's:		Correlated defaults	
11	A	1	0.452884	-0.03955	0.734847	0.069335	0.053538		0.670256		0	
12	B	0.452884	1	0.060151	0.051447	0.696404	0.462757		0.830776		0	
13	C	-0.03955	0.060151	1	0.022945	0.330949	0.444655		0.066109		0	
14	D	0.734847	0.051447	0.022945	1	-0.04023	-0.03106		0.795843		0	
15	E	0.069335	0.696404	0.330949	-0.04023	1	0.063667		0.68178		0	
16	F	0.053538	0.462757	0.444655	-0.03106	0.063667	1		0.501828		0	
17												
18									Single asset default			
19									0.1062			
20												
21									Six asset simultaneous default rate			
22									0			
23									0.0002			

FIGURE 5.11
Moving from the correlation matrix to a copula.

5.4 Third Example: How Correlated Are Home Prices?—Copulas

Just how correlated are home prices? City home prices5.xlsx provides quarterly data on home prices and changes in home prices from 1991 through 2010 in two American cities: New York and Las Vegas; the latter was at the heart of the subprime mortgage crisis.* We wish to investigate the correlation patterns. For example, consider the scatter plot of price changes in New York and Las Vegas shown in Figure 5.12.

There appears to be a clear pattern of positive correlation, but the correlation is much tighter at the low ends of the distributions and weaker in the middle and upper ends.† A Normal *copula* (which produces a symmetric and linear correlation) will not capture this pattern well. It is widely believed (at least now) that housing markets are more tightly coupled when prices are falling significantly than when the changes are more moderate.

* These data come from the Case-Shiller home price indices. Since these are time series data, we should use time series modeling techniques (Chapter 6) or analyze the changes in the prices, as we did with the State home price data. We use the indices directly here, without a time series approach, because this is a good illustration of the ability of the Empirical copula to reproduce correlation patterns closely.

† This pattern will not be as evident in an Excel scatter plot as in this ModelRisk scatter plot (produced through the Bivariate Copula Fit window) because the latter scales the data by percentiles of the distribution while the former uses the absolute scaling in the data (in this case, the percentage change in home prices).

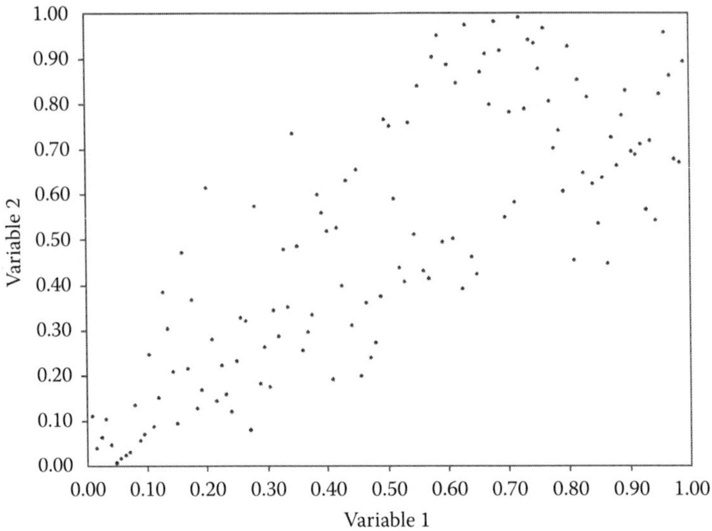

FIGURE 5.12
Home price changes in New York and Las Vegas.

There are a number of different copula structures available in the finance literature and in ModelRisk. Among these are correlation patterns (copulas) that can better represent this pattern than does the Normal copula. These patterns can be estimated from data or chosen by the analyst, just as with probability distributions (see Chapter 4). ModelRisk permits a number of copulas to be fitted to data; these are found under the Fit button on the ModelRisk toolbar. Selecting the Bivariate Copula Fit and choosing all five potential copulas gives us Figure 5.13.

Based on the ranking by the information criteria (SIC, AIC, and HQIC), the Clayton copula fits these data best, followed by the Gumbel copula. There are radio buttons for Chart mode that permit you to compare the data with the fitted copula (either viewed separately or simultaneously). If you explore these different views for each of the copula types, you will see that the Clayton copula indeed better captures the tighter relationship between the two cities' housing price changes when prices fall than does the Normal copula.

As with distribution fitting in ModelRisk, there are a variety of options available: to include parameter uncertainty in the fitted function and to output the copula, its parameters, or an object into your spreadsheet.

It is also possible to model the correlation patterns among more than two variables. State home prices5.xlsx provides quarterly data for four states at the heart of the housing crisis: Arizona, California, Nevada, and Florida.* It is

* The data are for the Housing Price Index published by the Federal Housing Finance Agency and an analysis of these data, using copulas, appears in Zimmer, D. The role of copulas in the housing crisis. *Review of Economics and Statistics*, forthcoming.

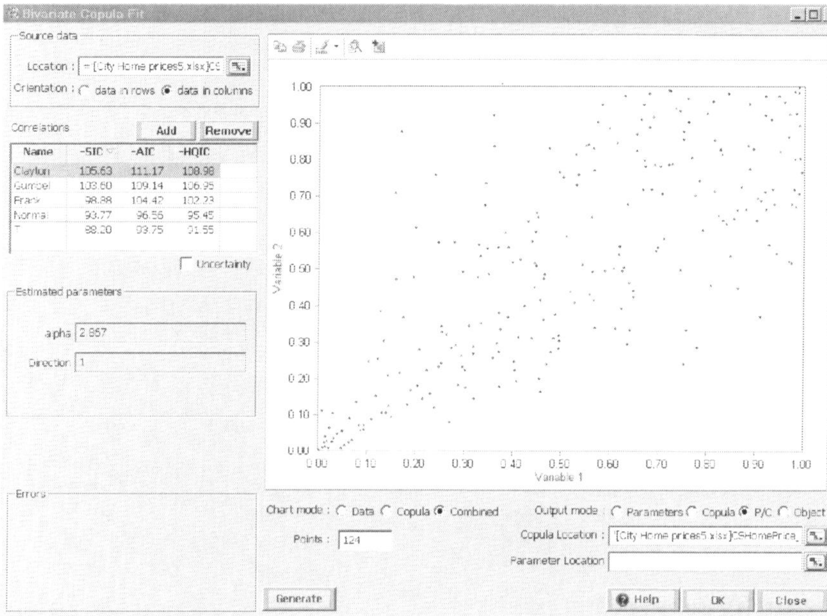

FIGURE 5.13
Fitting a copula to the home price change data.

possible to fit copulas to these series simultaneously, using the Multivariate Copula Fit, as shown in Figure 5.14.

In this case, the T copula best fits the correlation among the four data series. It is important to note that the Multivariate Copula Fit will only fit one copula pattern to all of the variables; other copulas may be better fits when considering any two individual states. Further, only the T and Normal Multivariate copulas permit the strength of the correlations to vary among the multiple variables (shown in the correlation matrix in Figure 5.14). The Gumbel, Frank, and Clayton copulas impose the restriction that the degree or "strength" of the correlation patterns be the same across all variables.

In the present case, we will use the Multivariate T copula.* We output the copula in row 161 of the spreadsheet. (Insert it into a 1 × 4 cell range because there are four copulas, one for each variable.) The four copula values are then used as U values in the distributions that we fit to each data series. The U values ensure that the historical correlation patterns are captured when

* There is a difficult choice to be made here. The T copula fits the multivariate data best and permits the strength of the correlations to vary among pairs of variables. Some pairs of states are fit better with different copulas; the Frank copula fits best for Nevada and Florida, but this copula is not the best overall fit and does not permit the correlation strengths to vary. A third option would be to use an Empirical copula, discussed in the next section.

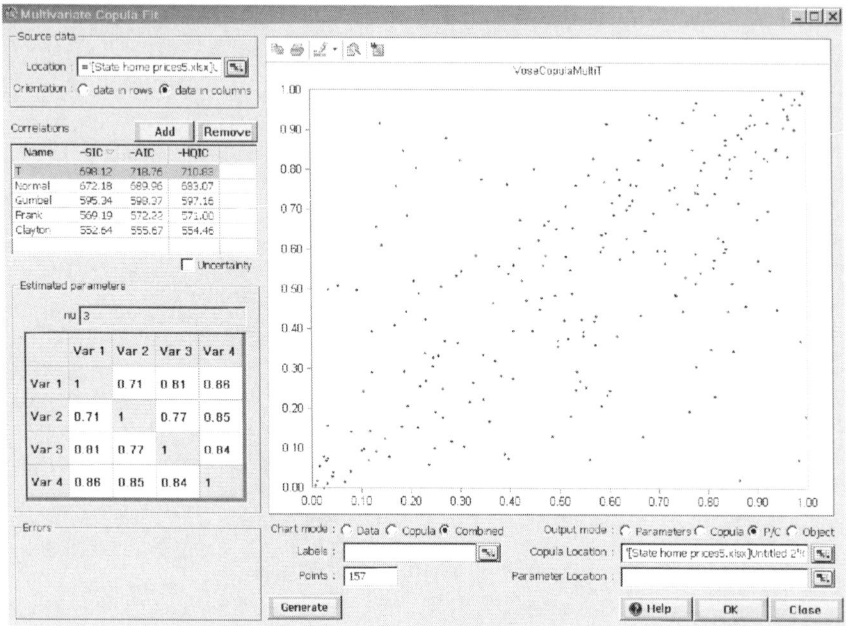

FIGURE 5.14
A Multivariate copula for home prices in four cities.

simulating each of the distributions. Figure 5.15 shows the completed section of the spreadsheet after fitting each of the distributions and adding the copula U values into each.*

We could now use these distributions in a simulation to model how housing prices change in the four states, with the correlation patterns in the changes in housing market prices captured with the copula.†

* Only the first series, column G, is fully expanded: You can see where G161 is added to the Cauchy distribution fit to capture the correlation. The two FALSE entries in the copula show that the data are in columns and that we have not chosen to include uncertainty in the fit.

† The current approach simulates the changes in price over time, which is a typical way of modeling parameters over time. An alternative approach would be to model this using a time series that takes into account what happened in the last one, two, three, or more time periods (or other structures, e.g., trends, volatility clustering, etc.) more than in the earlier data. Some time series methods are available to capture such structures, and this is the subject of Chapter 6.

	F	G	H	I	J
161	Copulas	=VoseCopulaMultiTFit(G3:J159,FALSE,FALSE)	=VoseCopulaMultiTFit(G3:J	=VoseCopulaMultiTFit(G3:.	=VoseCopulaMultiTFit(G3:J1
162	Distributions	=VoseCauchy(0.012499232858847,0.009112836666552,G161)	=VoseExtValueMin(0.0206:	=VoseCauchy(0.00841272:	=VoseCauchy(0.00995348:

FIGURE 5.15
Using the Multivariate copula in simulation.

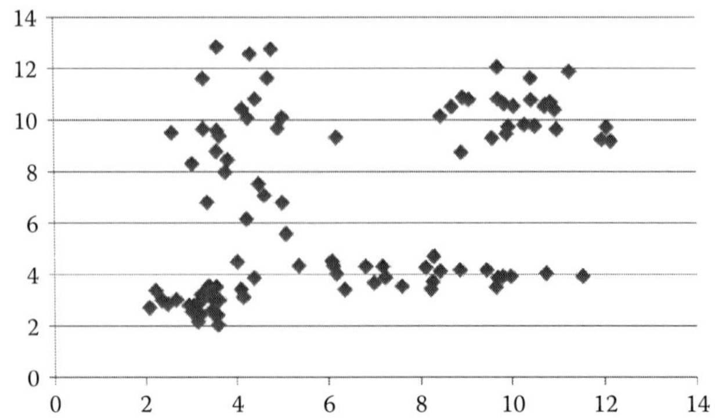

FIGURE 5.16
A non-normal copula pattern.

5.5 Empirical Copulas

There is an additional type of copula that is available in ModelRisk: an Empirical copula. This copula will mimic whatever pattern the data reflect—even highly complex nonlinear relationships that could not otherwise be captured in a simulation. Empirical5.xlsx contains hypothetical data for a sample of individual viewing habits for two cable television channels. Each point shows the viewing hours per month for an individual. Figure 5.16 shows a scatter plot of the data.

The program producer is considering bundling the two channels, and Figure 5.16 clearly suggests that there are four distinct market segments. The correlation pattern is clearly not normal and does not look much like the other copula patterns either. However, the Empirical copula (found under the Fit button on the ModelRisk toolbar) can reproduce this pattern quite closely, as shown in Figure 5.17. (Try changing the radio button in the Fit Empirical Copula dialog box between Data and Fitted Copula. Figure 5.17 shows the combined data points.)

Indeed, the Empirical copula accurately represents the original pattern in the data.* This can be seen more clearly if we fit distributions to each of the data series and then simulate new data using these distributions and the Empirical copula. The section of the spreadsheet designed to accomplish this is shown in Figure 5.18, and the results of the simulation are shown in Figure 5.19.

* The reason this scatter plot looks different from that in Figure 5.16 is again due to the different scaling in the ModelRisk window (showing percentiles of the distribution) and by Excel (showing the data values).

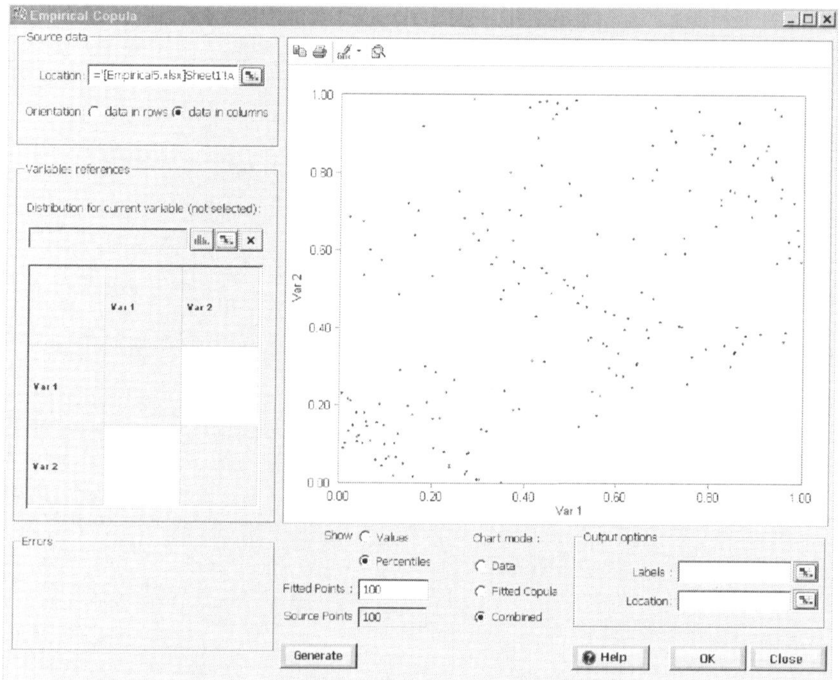

FIGURE 5.17
Fitting an Empirical copula.

	A	B	C
102	=VoseCopulaData(A2:B101,FALSE)	=VoseCopulaData(A2:B101,FALSE)	Copula
103	=VoseDataObject(A2:A101)	=VoseDataObject(B2:B101)	Data Object
104	=VoseBetaSubjFitObject(A103)	=VoseJohnsonBFitObject(B103)	Distribution Fit
105	=VoseOutput("Hours1")+VoseSimulate($A$104,$A$102)	=VoseOutput("Hours2")+VoseSimulate($B$104,$B$102)	Simulated data (with copula)

FIGURE 5.18
Simulating with the Empirical copula.

Figure 5.19 is adjusted to show 100 simulated points (to match the original 100 data points) using the same measurement scale as the original data. The results are quite close to the original data shown in Figure 5.16.

The Empirical copula is always an option for modeling correlation patterns (either in bivariate or multivariate data). The advantage is that, typically, it will more closely match the existing data than any other copula. When using Empirical copulas, there are two shortcomings to recognize, however. First, Empirical copulas are computationally intensive, so extensive use of them can slow down simulation models considerably. Second, and more importantly, it may not always be desirable to match the pattern in the data closely. After all, closely matching past data may be a poor way to forecast the future. It may be more accurate to represent the pattern in the

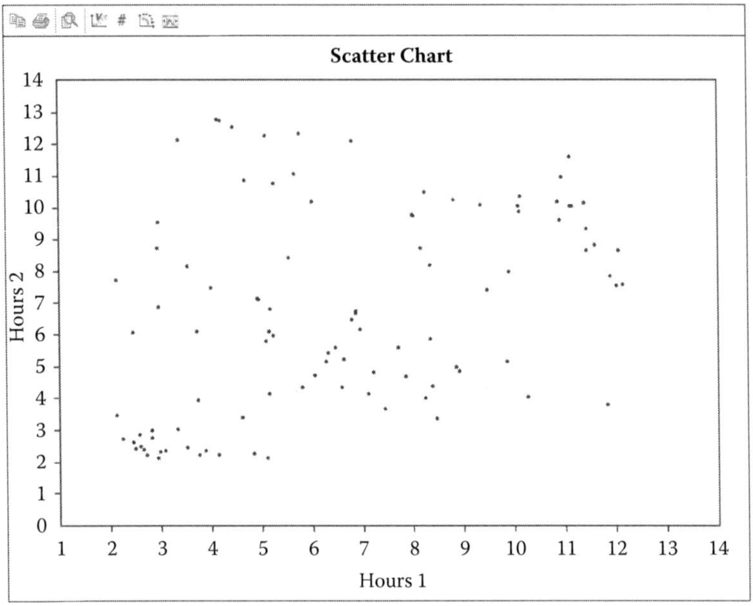

FIGURE 5.19
Empirical copula results.

data than to match one particular realization of the unobservable process that generated the data.

5.6 Fourth Example: Advertising Effectiveness

Suppose that a company has data on levels of advertising (thousands of dollars/month) and sales (thousands of dollars/month) from 10 different local markets (data in Table 5.1). The company wishes to estimate the effectiveness of its advertising and to predict the level of sales for a certain advertising budget. Taking a very simplistic approach, we could attempt to model the relationship between advertising and sales using a LOOKUP function. For an advertising budget of $7,500, this will produce estimated sales of $13,000 (if we do not look for an exact match because an exact match will produce an Excel error since there is no exact match in the table). This LOOKUP approach, however, ignores any uncertainties and does not interpolate between different advertising budgets.

TABLE 5.1

Hypothetical Advertising and Sales Data

Advertising	Sales
1	2
2	7
3	3
4	7
5	10
6	9
7	13
8	12
9	19
10	18

An alternative approach might be to estimate a copula to capture the relationship between the two variables. But this approach treats the two variables as randomly fluctuating, albeit with some dependence between them. While modeling a relationship using a copula approach does capture the uncertainty around the relationship, it assumes that both variables (i.e., advertising and sales) are uncertain. When one of the variables is not uncertain (in this case, advertising) or is based on another number of factors (e.g., another part of the model), a copula approach is typically not appropriate. In this case, we can estimate the "dependency" relationship using *regression analysis*.

There are many forms of regression analysis and Excel is capable of producing a variety of these; we will only consider linear regression here. Linear regression assumes that there is a linear relationship between a response variable, Y, and a number of explanatory variables, $X_1, \ldots, X_n$. In the present case, Y is sales, and we only have one X variable, advertising (i.e., a simple linear regression).*

* Note that linear regression can model nonlinear relationships between Y and X. If we define a second X variable to be X2, then modeling Y to depend on X and X2 is a linear model (in the two X variables) even though it is a nonlinear model of X alone. A very common approach is to model the logarithm of Y as a linear function of the logarithm of X, which is a nonlinear function of Y and X (but linear in terms of the logarithms).

5.7 Regression Modeling

Linear regression does impose a few restrictions on the relationship between *Y* and *X*. The assumptions that underlie linear regression models are the following*:

- The data come from a sufficiently large *random* sample.
- The residuals (the difference between what the model predicts and the actual *Y* values) are normally distributed.
- The distributions of these residuals are identical for different values of the *X* variables.

Regression models are quite robust to deviations from these assumptions, however, and are commonly used to capture relationships from real data that do not strictly satisfy these conditions. Also, refinements of the standard regression model are available to deal with many common departures from these assumptions.†

Simple linear regression models are based on finding the best fitting linear model between the *Y* and *X* variables. "Best fitting" is defined as the linear model that minimizes the sum of the squared deviations of the actual data points from the estimated linear relationship. Figure 5.20 illustrates the procedure for our hypothetical advertising–sales data.‡

Least squares regression can be thought of by considering many potential straight lines that can be drawn through the data points and finding the one that minimizes the sum of the squared deviations of the points from each line. The arrow labeled "unexplained" in Figure 5.20 represents the deviation of that particular data point (advertising = 9, sales = 19) from the straight line $Y = 1.76 + .33X$, which is the actual least squares regression line (i.e., the best fitting line) for these data. The arrow labeled "explained" represents the deviation of this data point from the average sales value in the data (= 10 in these data). It is explained because our model assumes that sales are dependent on the advertising level, and the advertising level at that point is greater than the average value in the data (9 > 5.4). Since the advertising level at this point is relatively high, our least squares regression model predicts (or explains) that the resulting sales will be higher than average (predicting a

---

* Readers are urged to refer to any standard statistics book if they require further understanding of the statistical theory behind regression models or detailed information for conducting such analysis. Our focus is on implementing simple regression models in Excel and using Monte Carlo simulation to represent the uncertainty inherent in such models.

† For example, when the variance of the residuals does vary with the levels of X, weighted regression can often be used to overcome this problem. This technique is covered in any good intermediate statistics text and can easily be implemented in Excel.

‡ Figure 5.20 is obtained by using a scatter chart from the Insert menu in Excel and then adding a Trendline by right-clicking on a point in the chart and selecting "add Trendline."

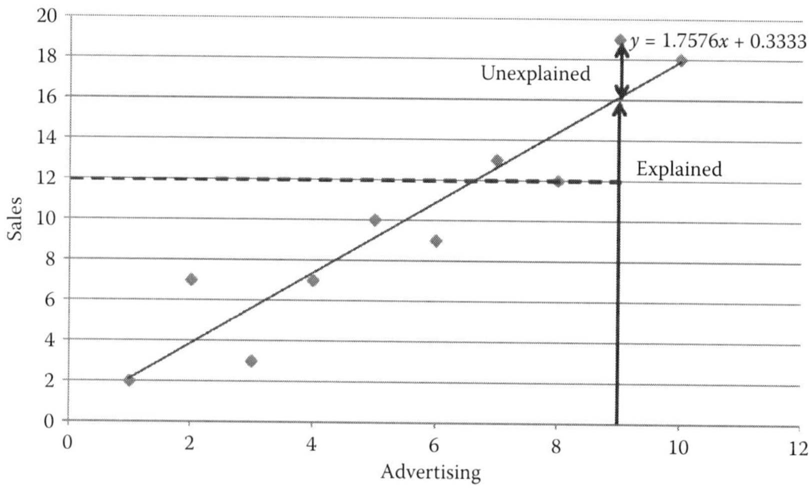

FIGURE 5.20
The least squares regression model.

sales level of 16). It does not predict, however, why the sales are as high as 19. The remaining \$3,000 of monthly sales (\$19,000 minus \$16,000) is unexplained by this model. It is the sum of all these unexplained squared deviations that the least squares regression line minimizes.

Excel provides a number of ways to obtain the least squares regression line. The most comprehensive solution is obtained by choosing Data Analysis on the Data menu.* A variety of statistical methods are available within Data Analysis; choose Regression from the drop-down list. Insert the sales data as the *Y* range and the advertising data as the *X* range. (Click Labels and include the labels in the ranges.) You will get the output shown in Figure 5.21.

The Regression Statistics section provides some overall information and statistics about the relationship between *X* and *Y*.

Multiple R is a number between –1 and +1 that measures the quality of the linear regression. (For a simple linear regression, this is the same as the Pearson correlation coefficient, which you can calculate using Excel's CORREL function.) *R-Square,* its squared value, measures the extent to which the independent variables account for the variation in the response variable. R-Square ranges from zero to one; a value of one means that every predicted value is exactly equal to its actual value. This is equivalent to the least squares regression line passing exactly through every actual data point. The *Adjusted R-Square* makes an adjustment for the number of independent variables (in this case, just one—the advertising level). R-Square will always

* Data Analysis is an Excel add-in that may not be activated when you first install Excel. If it does not appear on the Data menu on the toolbar, go to the Excel start button, click on Excel Options and then Add-Ins, and choose "Go" for Manage Excel Add-ins. There is a check box for Analysis ToolPak that you should check.

	A	B	C	D	E	F	G
1	SUMMARY OUTPUT						
2							
3	*Regression Statistics*						
4	Multiple R	0.93743687					
5	R Square	0.87878788					
6	Adjusted R Square	0.86363636					
7	Standard Error	2.09617256					
8	Observations	10					
9							
10	ANOVA						
11		*df*	*SS*	*MS*	*F*	*Significance F*	
12	Regression	1	254.8484848	254.8485	58	6.21252E-05	
13	Residual	8	35.15151515	4.393939			
14	Total	9	290				
15							
16		*Coefficients*	*Standard Error*	*t Stat*	*P-value*	*Lower 95%*	*Upper 95%*
17	Intercept	0.33333333	1.431958467	0.232781	0.8217762	-2.96876881	3.63543548
18	advertising	1.75757576	0.230781003	7.615773	6.213E-05	1.22539381	2.2897577

FIGURE 5.21
Regression model results.

increase if more independent variables are added to the model, but the Adjusted R-Square will only increase when statistically meaningful independent variables are included. The *Standard Error* is the standard deviation of the residuals from the model. *Observations* are the number of data points.

The ANOVA (Analysis of Variance) table also provides measures of the overall fit of the model to the data. It does not, however, add much to the interpretation of the model beyond what is in the other sections of the regression output. Briefly, it decomposes the *Total* sum of squared residuals in the regression into the fraction that is accounted for by the regression model (*Regression*) and that which is not (*Residual*). It divides these sums of squared residuals (*SS*) by the appropriate degrees of freedom (*df*—a function of the number of data points and the number of independent variables) to get the mean square (*MS*) and then gives the probability of getting a mean square ratio larger than this value if there was no relationship between the two variables, using an F distribution. The resulting *Significance F* is an overall measure of the significance of the independent variables. (It can be interpreted as the probability of obtaining an F statistic at least as large as the one observed, given that the null hypothesis of no dependency between advertising and sales is true.) If there were no relationship between the independent and dependent variables, then the regression sum of squares would likely be low, so the MS and F values would be low, and the Significance F would only rarely be less than 5%. Conversely, the stronger the dependency is, the larger the regression SS relative to the Residual SS, the greater the F value, and the smaller the Significance F will be. If this value is sufficiently low, then the probability that the independent variables could account for this much of the variation in the dependent variable is low enough to conclude that the overall relationship is significant and not just the result of chance.

The last table in Figure 5.21 provides information about the variables used in the model. The *Intercept* is part of the linear relationship, but it represents all of the variables that were not included in the model, so it is not generally interpretable. There are rows for each of the independent variables; in this case, only advertising appears as an independent variable. The *Coefficient* represents the impact of a unit increase in the independent variable upon the response variable, accounting for all other variables in the model. In this case, spending an additional $1,000 on advertising per month would lead to an expected additional $1,758 in monthly sales.

Statistical theory (as well as some logical thinking) tells us that if we took a different random sample (e.g., we redo our survey and by chance get different observations), the data points and associated least squares regression line would look a bit different.* The *Standard Error* represents the standard deviation for the values of this Coefficient with repeated random samples. Dividing the Coefficient by this Standard Error gives the t statistic (*t Stat*), which measures how many standard deviations away from zero the Coefficient is. So, if we were to hypothesize that the real effect of the variable is zero (i.e., the Coefficient of the true model is zero, and the only reason our samples provide an estimate different from zero is due to random sampling error), then we can ask how likely it is that a random sample could produce an effect (Coefficient) this large. The answer to that question is the *P-value.* It tells us the probability that chance (i.e., random sampling) alone could produce an estimated effect this strong, under the assumption that the true effect is zero. If this P-value is sufficiently low, then we would reject this hypothesis and say that we have evidence that the effect of the variable is statistically significant.

An alternative to the hypothesis test and P-value, with an interpretation that is often easy to understand, is provided by the *Lower 95%* and *Upper 95%* calculations.† These provide a 95% confidence interval for the size of the Coefficient, based on our sample data. Thus, we can say we are 95% confident that the true impact of a $1,000 increase in monthly advertising will lead to an increase in monthly sales that is between $1,225 and $2,290.

The size of the confidence interval is calculated by adding or subtracting an appropriate number of Standard Errors from the estimated Coefficient. The better the fit is, the lower this Standard Error will be and the narrower the resulting 95% confidence interval for the Coefficient will be. A better fit to the data means there is less uncertainty about the true parameters of the regression model.

The P-value and 95% confidence interval are intimately related: They are both based on the Standard Error and estimated Coefficient. If the 95%

* The larger our sample is, the less this difference will likely be. This is the reason why, from a statistical perspective, larger samples are typically preferred to smaller ones.

† If you produce this display, you will note that these 95% columns appear twice. That has been the case for every version of Excel. We do not know why these columns are duplicated and no information is provided by the repeat.

confidence interval does not contain zero, then we are at least 95% confident that the variable is either positively or negatively related to the response variable. (In this case, there is a positive association between advertising and sales.) The corresponding P-value will be less than 5%. Whenever the 95% confidence interval includes the value zero, then the P-value will be greater than 5%, and we cannot say that there is a significant association between the independent and dependent variables.

Use of 5% at the critical P-value for determining statistical significance is fairly standard academic practice, although other values are sometimes used in particular applications. As a practical matter, however, the choice of the threshold P-value is a *judgment* that the analyst must make. The choice of a low threshold for the P-value (say, 1%) places a high burden of proof for declaring the effect significant. But this means that there is a correspondingly larger risk that we will not find a statistically significant relationship even if one really exists. Conversely, declaring an effect statistically significant at a higher threshold value (e.g., 10%) reduces the risk of failing to acknowledge a relationship in the data, but runs a higher risk (up to 10%) of declaring an effect significant when it really is only the result of random sampling. There is always a trade-off between these two types of potential errors (called Type I and Type II errors), and the choice of a critical P-value (alternatively, the choice of the confidence interval, e.g., 90, 95, 99%) should result from an understanding of the relative costs of these two types of errors.

For example, in studies concerning public health, many experts consider the use of a higher P-value to be appropriate. They argue that the cost of failing to recognize a significant effect of a potential hazard on human health is greater than the cost of declaring an effect significant when it really is not. Of course, it is easy to imagine cases where the reverse could be true: A particular potential hazard could be of little consequence but it might be very costly to mitigate the potential impact. Rather than arbitrarily choosing a P-value threshold, we advise you to think carefully about the relative risks of Type I and Type II errors in the particular issue you are analyzing. Further, the choice need not be made in a simulation model because the model takes into account the degree of uncertainty regarding the relationship. We will now show how this is done.

5.8 Simulation within Regression Models

Returning to our advertising effectiveness case, Figure 5.21 suggests that the regression model may be a good description of how advertising influences sales. The R-square is relatively high, the advertising coefficient is highly statistically significant (P-value = .000006), and the residuals appear to be randomly scattered around the regression line.* Suppose that we wish to forecast the sales level that would result from a monthly advertising level of $7,500. We could use the estimated relationship, sales = 1.76 + .33*advertising, to derive the sales forecast. But because of the limited amount of data and because of chance, there is uncertainty about the true relationship between advertising and sales. Fortunately, the regression analysis output provides evidence regarding the level of uncertainty in the relationship. Since we have the standard deviations for the intercept (1.76 with standard error = 1.43) and slope (.33 with standard error = .23) and the regression assumptions imply that the true intercept and slope will follow a Normal distribution around the sample estimates, we have the information required to simulate the possible intercepts and slopes of the true advertising–sales relationship.

There is one complication, however. The uncertainty distributions of the intercept and slope are not independent; as the intercept varies, the slope of the least squares regression line through the original data points will depend on this intercept. If an intercept higher than 1.76 were simulated, we would find that the least squares regression line through the data points would need to be flatter, resulting in a slope lower than .33. Conversely, if we simulate a lower intercept than 1.76, the line will tend to be steeper (i.e., have a higher slope).

To simulate the uncertainty in the relationship correctly, we need to adopt a different procedure than simply simulating the intercept and slopes independently. The method we use is a *parametric Bootstrap:* Viewing the current data as one random sample of data points from the underlying process, we will generate additional random samples of new data and estimate the least squares regression model for each of these samples (also known as Bootstrap samples). The resulting simulation incorporates the full uncertainty inherent in our data.

Regression Model5.xlsx contains this parametric Bootstrap model for our advertising–sales data. Generating random samples of sales data relies on the use of the standard error of the residuals around the regression line. This is shown in Figure 5.21 as the Standard Error in the Regression Statistics table (2.09 in this case). It can also be found by using the STEYX function in Excel. Figure 5.22 shows the simulation model.

The first model (in the box in rows 14 through 20) simply finds the least squares regression line through the original data and then uses the regression standard error to simulate the potential sales that would result from an advertising level = 7.5. The bottom box (rows 22 through 28) estimates the least squares regression line based on the Bootstrap sample in F2:F12.

* Further diagnostics for the residuals are available and are recommended before settling on a final regression model. In the present case, Figure 5.20 provides a quick visual check that suggests that the assumptions of the linear regression model may be reasonably satisfied with these data.

	B	C	D	E	F
1					
2	Data:	advertising	sales		Bootstrap Sample
3		1	2		=$C$16+C3*$C$17+VoseNormal(0,$C$18)
4		2	7		=$C$16+C4*$C$17+VoseNormal(0,$C$18)
5		3	3		=$C$16+C5*$C$17+VoseNormal(0,$C$18)
6		4	7		=$C$16+C6*$C$17+VoseNormal(0,$C$18)
7		5	10		=$C$16+C7*$C$17+VoseNormal(0,$C$18)
8		6	9		=$C$16+C8*$C$17+VoseNormal(0,$C$18)
9		7	13		=$C$16+C9*$C$17+VoseNormal(0,$C$18)
10		8	12		=$C$16+C10*$C$17+VoseNormal(0,$C$18)
11		9	19		=$C$16+C11*$C$17+VoseNormal(0,$C$18)
12		10	18		=$C$16+C12*$C$17+VoseNormal(0,$C$18)
13					
14	Forecast **without** parameter uncertainty				
15					
16	Intercept	=INTERCEPT(D3:D12,C3:C12)			
17	Slope	=SLOPE(D3:D12,C3:C12)			
18	SE_xy	=STEYX(D3:D12,C3:C12)			
19		x	y		
20	Forecast	7.5	=C16+C20*C17+VoseNormal(0,C18)		
21					
22	Forecast **with** parameter uncertainty:				
23					
24	Intercept	=VoseOutput(ModelRisk!B24)+INTERCEPT(F3:F12,C3:C12)			
25	Slope	=VoseOutput(ModelRisk!B25)+SLOPE(F3:F12,C3:C12)			
26	SE_xy	=STEYX(F3:F12,C3:C12)			
27		x	y		
28	Forecast	7.5	=C24+C28*C25+VoseNormal(0,C26)		

FIGURE 5.22
Simulating a regression model.

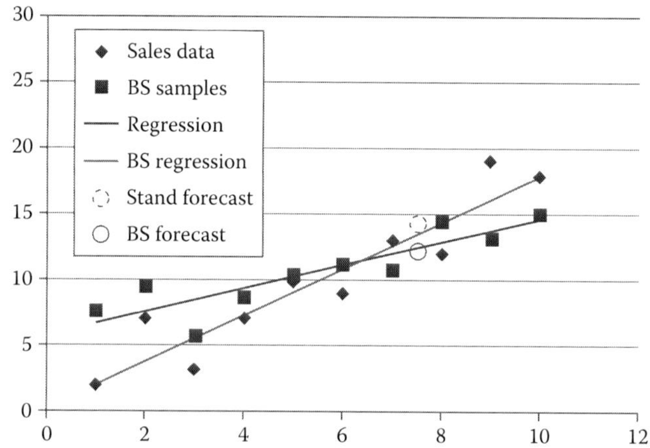

FIGURE 5.23
Illustrating the Bootstrap method.

(That is, in this box, the uncertainty around the regression line is taken into account.) The Bootstrap samples are constructed by simulating the sales level for each advertising level in the original data set. The standard error of regression is used (in a Normal distribution) to simulate a new sales level, based on the average level from the original regression line.* Each Bootstrap sample is then used to estimate the least squares regression intercept (cell C24), slope (cell C25), and standard error of regression (cell C26). These three parameters from fitting a regression to the Bootstrap sample are then used to simulate the sales forecast corresponding to the advertising level of 7.5 (cell D28).

Figure 5.23 illustrates the original data points and regression line, Bootstrap sample data points and corresponding regression line, and the predicted sales for each model that are associated with advertising = 7.5.

Each simulation (obtained by pressing the F9 key) will refresh the picture for a new Bootstrap sample and sales forecast. Running the simulation provides the forecast results shown in Figure 5.24.

The box plots reveal that the Bootstrap forecast is slightly more uncertain than what results from just simulating the sales level around the static

* The resulting prediction interval we obtain will be slightly narrower than what you would obtain from a standard statistics package. Theoretical prediction intervals are based on both X and Y being randomly sampled. (It is possible to build the correct formula in Excel for this classical prediction interval, but it is more complex than using the Bootstrap procedure shown here.) In the present case, we are not viewing the advertising levels as forming a random sample of advertising; these levels were chosen for particular reasons. Our assumption is that the resulting sales levels are variable and we are observing but one random sample of what the sales might be for each level of advertising. If both variables were the result of a properly conducted random experiment, then more complicated procedures would be required to run a proper simulation.

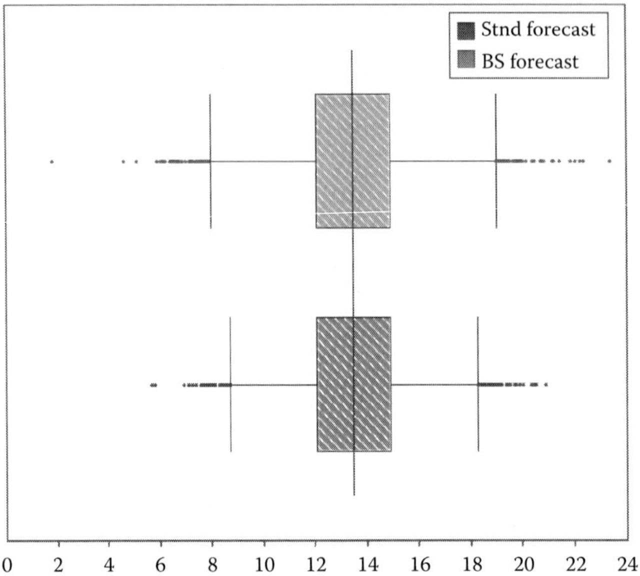

FIGURE 5.24
Box plot results of the Bootstrap procedure.

least squares regression line. This can be seen by the somewhat larger range shown on the box plot. The statistics, shown in Figure 5.25, show that the mean forecast sales level is almost the same, but that the standard deviation of the Bootstrap forecast is greater than for the standard forecast. The slightly higher uncertainty in the sales level forecast with the Bootstrap approach is not a surprise, given that it takes into account an additional level of uncertainty.

Options	BS Forecast	Stnd Forecast
Location		
Mean	13.522228	13.526939
Minimum	1.8606715	5.6998597
Maximum	23.408741	21.337069
# of Errors	0	0
# of Filtered	0	0
Spread		
St. dev.	2.2520487	2.080163254
Variance	5.0717232	4.3270792
CofV	0.16654422	0.15377931
Shape		
Skewness	0.001147746	0.0019118344
Kurtosis	3.6012678	2.9325064

FIGURE 5.25
Statistics for results of the Bootstrap procedure.

In the present case, the additional amount of uncertainty using the Bootstrap procedure is relatively small. It can, however, make an important difference, especially in situations with relatively few data. Recognizing the uncertain linear relationship produces an 80% confidence interval for the sales prediction ranging from around 10.9 to 16.2.*

5.9 Multiple Regression Models

The methods used in the Advertising–Sales model are readily generalized to models that have a number of independent X variables. We will not provide a complete analysis here, but in order to illustrate the appropriate Excel functions, consider the data and analysis shown in Figure 5.26.

Column A provides hypothetical output data (for example, number of packages sorted per hour at various sorting plants) and data on three inputs (for example, mechanized sorting lines, number of human sorters, and number of truckers). Use of Regression Analysis from the Data Analysis in Excel produces the nicely formatted SUMMARY OUTPUT shown in Figure 5.26. Careful inspection of the output shows that the model provides a good fit to the data and that all three inputs are statistically significant variables. (Note that Excel will only permit multiple independent variables to be entered into a regression analysis if they are in contiguous columns.) The Bootstrap procedure illustrated in Section 5.7 can now be used to simulate the output resulting from any particular combination of the three inputs. However, we cannot use the INTERCEPT, SLOPE, and STEYX functions because they only apply when there is a single independent variable. Use of the Regression Analysis capability in Excel is problematic since each Bootstrap sample is different and we would need to invoke Regression Analysis each time a different sample is generated. This is because Regression Analysis does not automatically recalculate if any of the data change.

Fortunately, Excel provides the LINEST function, which does recalculate as the data change and produces the required regression model output. However, it is not formatted very nicely. Figure 5.26 shows the function and its output (in cells F22:I26).

LINEST is an array function: The user must first highlight the required number of cells that the output will occupy. (It is five rows of data, and the number of columns equals the number of independent variables plus one.) The first argument in the function asks for the range for the

* The 95% confidence interval resulting from the Bootstrap forecast is (9.11, 17.97), while the standard forecast produces an interval of (9.42, 17.56). The classical prediction interval (if both X and Y are random samples derived from a statistical software package) is (8.33, 18.70). The latter is not appropriate here since X (advertising) is not random.

F22 {=LINEST(A2:A31,B2:D31,TRUE,TRUE)}

	A	B	C	D	E	F	G	H	I	J	K	L
1	output	input A	input B	input C		SUMMARY OUTPUT						
2	7668	4	27	16								
3	8204	4	27	17		*Regression Statistics*						
4	11382	6	31	22		Multiple R	0.942359					
5	6666	3	25	13		R Square	0.88804					
6	12328	7	38	16		Adjusted R S	0.875121					
7	17457	8	47	21		Standard Err	1005.561					
8	10016	4	29	16		Observatior	30					
9	11581	6	40	26								
10	8628	5	31	14		ANOVA						
11	11774	6	32	23			*df*	*SS*	*MS*	*F*	*ignificance F*	
12	13265	5	54	29		Regression	3	2.09E+08	69508445	68.74179	1.72E-12	
13	10918	5	31	26		Residual	26	26289968	1011153			
14	9024	5	33	22		Total	29	2.35E+08				
15	7998	3	25	19								
16	10648	7	34	17			*Coefficients*	*andard Err*	*t Stat*	*P-value*	*Lower 95%*	*Upper 95%*
17	10373	5	34	18		Intercept	739.1736	797.0019	0.927443	0.362227	-899.087	2377.435
18	11003	5	35	24		input A	825.7106	173.2512	4.765973	6.23E-05	469.5876	1181.833
19	14951	8	38	31		input B	94.77352	37.88778	2.501427	0.018996	16.89407	172.653
20	12941	7	39	22		input C	117.9759	38.03784	3.101539	0.004593	39.78796	196.1637
21	13327	6	32	27								
22	7576	3	34	11		117.97585	94.77352	825.7106	739.1736			
23	7013	4	28	6		38.037837	37.88778	173.2512	797.0019			
24	8300	3	26	18		0.8880398	1005.561	#N/A	#N/A			
25	6400	3	19	13		68.741794	26	#N/A	#N/A			
26	14465	8	44	31		208525335	26289968	#N/A	#N/A			
27	6802	3	24	8								
28	9992	4	28	25								
29	6568	3	25	8								
30	7539	3	21	13								
31	12314	8	43	24								

FIGURE 5.26
Multiple linear regression model.

response variable and the second argument contains the range for the independent variables; with multiple X variables, they must be in a contiguous range. The third argument (TRUE) tells Excel to include an intercept, and the fourth argument (TRUE) tells Excel to produce the full set of regression outputs. **Entering the array formula requires simultaneously selecting Ctrl + Shift + Enter.** The resulting table is unlabeled, but you will find that the relevant data required for our regression simulation are contained in the LINEST function output. In particular, the top row provides the regression coefficients. (Note that the coefficients are in reverse order, with the constant being the last value.) The second row shows the standard errors associated with these coefficients, and the third row gives R-square and the Standard Error, with additional regression data on the fourth and fifth rows.

We can now use the Bootstrap procedure, relying on the Standard Error and regression coefficients from the LINEST function. The LINEST function will automatically recalculate if any of the input data are changed. Thus, LINEST is useful in simulation models since simulations will usually involve the data changing with each sample. Users are recommended to insert their own text labels next to the LINEST output to remind them of what each number represents.

To implement the Bootstrap procedure, the following four steps are required:

1. Use the Standard Error from the original least squares model to generate Bootstrap samples. (Use the coefficients of the regression model and add the random term of a normal distribution with mean 0 and standard deviation = Standard Error.)
2. Use a second LINEST function to estimate the regression model for each Bootstrap sample.
3. Use the coefficients and Standard Error from this LINEST function to estimate the Y value for the combination of X variables in which you may be interested.
4. Run a simulation to model the uncertainty inherent in the least squares regression model.

Within the simulation model, every iteration will now generate Bootstrap samples, reestimate the regression coefficients, and generate a predicted value of the dependent variable. Thus, running the model for 10,000 iterations provides you with the uncertainty around the value of the dependent variable (e.g., number of packages sorted per hour at various sorting plants).

5.10 The Envelope Method*

Expert opinion is commonly used in Monte Carlo simulation, and often expert opinion is used to model relationships between variables. We return to the Apple stock valuation example from Chapter 4. To illustrate the use of the Envelope method, we will focus on one of our experts (Expert A). Expert A provided estimates for terminal revenue growth of 3, 5, and 6.5% (minimum, most likely, and maximum) and estimates for Beta of 1.35, 1.4, and 1.5. This information was solicited as independent data but these two variables are in reality likely to be related. Terminal growth rates are usually assumed to equal the long-run expected growth rate for the economy, which is generally thought to be around 3%. Sustained growth rates above this level are typically not realistic and should be associated with increased risk—in other words, with higher values for Beta.

Subsequently, we asked Expert A to provide information about the potential relationship between terminal growth rates and values for Beta. For each of the terminal growth rates, we solicited a minimum, most likely, and

* This section is more advanced and can be skipped without interfering with the continuity of the text.

TABLE 5.2

Expert Opinion about Relationship between Terminal Growth and Beta

For each terminal growth rate below, provide an estimate for Beta	Minimum Beta	Most Likely Beta	Maximum Beta
3%	1.32	1.35	1.39
5%	1.4	1.45	1.52
6.5%	1.52	1.6	1.7

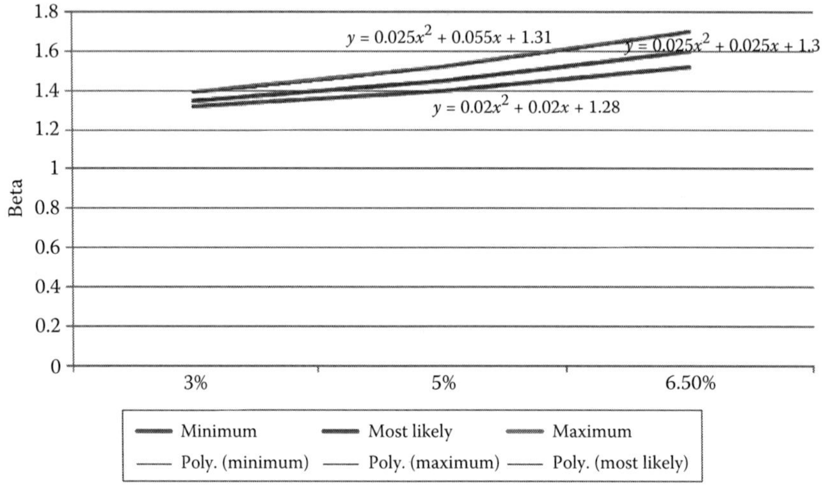

FIGURE 5.27
Fitted Trendlines.

maximum value for Beta that Expert A felt would be appropriate. Table 5.2 shows the estimates we obtained.

This table is reproduced in cells K7:N10 of spreadsheet Valuation5.xlsx. The Envelope method uses this information and regression analysis to estimate the relationship between the two variables. We can estimate the regression relationship using the Excel Data Analysis Regression tool, the LINEST function, or, as we do in this case, by fitting a Trendline to these data. We fit a second order polynomial in order to get a perfect fit through the data in Table 5.2.* The fitted Trendlines are shown in Figure 5.27.

* Since we are soliciting expert opinion, we are not interested in modeling the uncertainty about the regression fit; therefore, we eliminate this uncertainty by choosing a quadratic that will fit perfectly through the three data points for the minimum, most likely, and maximum values for Beta.

	A	B	C	D	E	F	G
1	AAPL						
2	Parameters		Expert	min	most likely	max	
3	Sales growth	0.3332					=B3
4	COGS	0.65					PERT Objects
5	annual cost decrease	=VoseSimulate(G5)	A	0	0.01	0.02	=VosePERTObject(D5,E5,F5)
6							
7							
8	R&D %	=VoseSimulate(G8)	A	0.02	0.0275	0.03	=VosePERTObject(D8,E8,F8)
9							
10							
11	SG&A%	0.15					
12	terminal growth	=VoseSimulate(G12)	A	0.03	0.05	0.065	=VosePERTObject(D12,E12,F12)
13							
14				using trendline equations (functions of B12)			
15	Beta	=VoseSimulate(G15)	A	=1.28+2*B12+2.5*B12^2	=1.3+2.5*B12+2.5*B12^2	=1.31+5.5*B12+2.5*B12^2	=VosePERTObject(D15,E15,F15)

FIGURE 5.28
Valuation model showing the Envelope method.

The Envelope method proceeds by simulating the terminal growth rate (using the original PERT distribution obtained from Expert A), then simulating Beta using a PERT distribution with the minimum, most likely, and maximum values derived from these Trendlines. This section of the revised model is shown in Figure 5.28.*

The Trendline equations for the minimum, most likely, and maximum Beta values are in cells D15:F15. Each simulated terminal growth rate in B12 will impact these parameters of the PERT distribution (Object in cell G15), which is used to simulate the value of Beta (in cell B15). In this way, the potential higher values for terminal growth will be associated with higher values for Beta—with the result that the (higher) future profits will be more heavily discounted.

Running the simulation gives the Apple valuation shown in Figure 5.29. The mean valuation is around $203 billion, with an 80% confidence interval ranging from $178 billion to $227 billion. This should be compared with the results if we used Expert A's earlier input, which provided PERT parameters for Beta that were independent of the terminal growth rate. If we use those PERT parameters and run the simulation, we obtain an expected company value of $212 billion, with an 80% confidence interval ranging from $185 billion to $240 billion. Evidently, accounting for the relationship between terminal growth and company risk leads to decreased overall valuation.† The Envelope method provides a means for capturing important dependencies in expert opinion modeling.

* Note that we have adjusted the Trendline equation by multiplying the coefficients for the X terms by 100 to reflect the fact that the spreadsheet measures the terminal growth rate as a percentage.

† It also decreases the uncertainty about valuation somewhat. This is due to the fact that the more optimistic views for terminal growth are partially offset by higher discounting of future profits (and vice versa), thereby reducing the range of potential overall value.

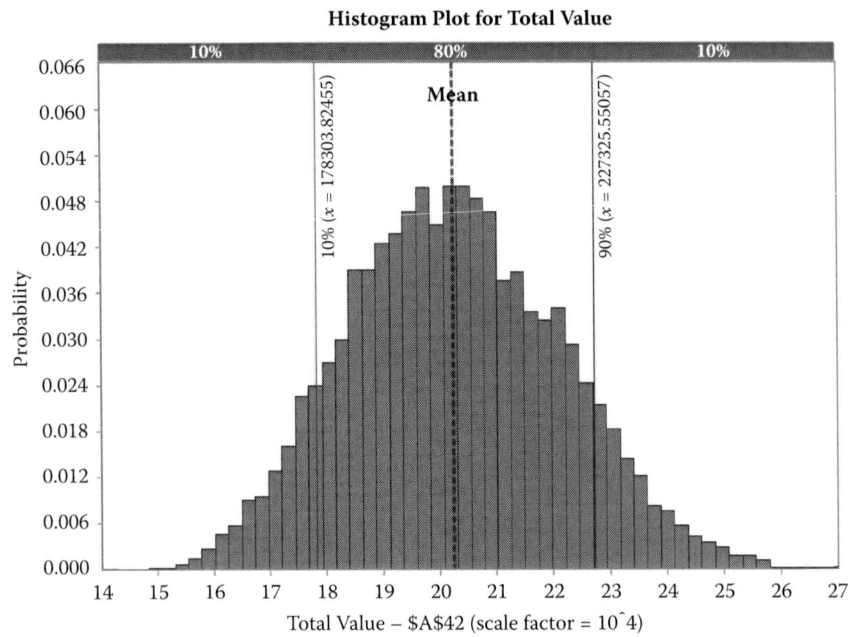

FIGURE 5.29
Apple valuation results using the Envelope method.

5.11 Summary

We have explored a variety of ways to model and simulate relationships among variables. As a general rule, it is always best to use the simplest model that captures the essential nature of the relationships. Logical statements are powerful and simple to use and can capture many relationships where a relatively small number of values for a variable are involved (for example, where one variable takes on the value one if some condition for another variable is met, such as a drug is approved if greater than 20% of the treatments are successful).

When the number of values is too large to capture with logical statements, correlation and copulas may capture the interdependence. Linear correlation is easiest and permits varying strengths in the relationships between variables to be modeled. More complex copulas should be used when Normal copulas inadequately reflect the data.

If a causal relationship is suspected in the data, then regression modeling is superior to modeling the correlation between variables. Copulas permit dependence, but assume that the variables are varying simultaneously, according to some uncertain pattern. Regression assumes that some set of variables can be used to predict linearly a variable of interest. Regression models not only provide a best fitting relationship, but also provide measures of the uncertainty about its strength. Simulation can then be used to model this uncertain relationship between variables.

CHAPTER 5
Exercises

5.1 DRUG DEVELOPMENT

Consider a pharmaceutical drug for treating Type 1 diabetes that is ready to enter phase 1 of testing for FDA approval. Initially, there is a 60% chance that the drug will be effective on any individual patient. Phase 1 is expected to cost $5 million and will last 1 year; 100 volunteers will be tested and the drug will successfully complete phase 1 if more than 50% of patients are successfully treated. (Assume that the tests on each individual are independent.) In phase 2, 250 volunteers will be tested over a 2-year period and this will cost $10 million. The probability that the drug will be successful in treating Type 1 diabetes if it passed phase 1 will be 70% for each patient. The drug will successfully complete phase 2 if at least 60% of the patients tested are successfully treated for Type 1 diabetes.

In phase 3, testing expands to 5,000 people for longer term testing and will take 4 years and cost $25 million. If the drug has successfully passed phase 2, the probability that it will successfully treat each patient for Type 1 diabetes is 80%. The drug will be FDA approved in stage three if it successfully treats Type 1 diabetes in at least 80% of patients. If approved, the cash flows will be $30 million per year for 15 years for the Type 1 diabetes drug if it proves effective for the entire Type 1 population.

a. Assuming a cost of capital of 10%, what is the expected net present value of this drug development? What is the probability that developing this drug will be profitable?
b. Estimate the NPV and the probability that drug development is profitable assuming that each stage of the development process is successful. Explain the results.

5.2 AIRFARES

Exercise5-2.xlsx contains data for all continental U.S. domestic air routes that average at least 10 passengers per day.* Data are provided for the originating and terminating city (and airport code) for each route, along with the average one-way distance, average number of passengers per day, market share of the largest carrier, code for the largest carrier, and the average fare charged on the route.

* The data come from a random sample of 10% of all airline tickets published quarterly by the Office of Aviation Analysis, US Department of Transportation. These data are the first-quarter 2010 Domestic Airline Fares Consumer Report.

a. Build a multiple regression model explaining airfares as a function of distance, passengers, and market share of the largest carrier. Interpret the meaning and statistical significance of each coefficient.
b. Suppose that, due to consolidation in the airline industry, we expect that the market shares of the largest carriers will increase on average by 10% and that the average number of passengers per day will increase by two (due to better scheduling and flight connections). Provide a point estimate of the impact on the average of the airfares across all routes that would result, using your regression model. Simulate the average fare across all routes without accounting for parameter uncertainty. (*Hint:* following Section 5.7 use the standard error of the whole regression for this. Use the fact that the regression model goes through the means of all the variables; that is, the average fare as a function of average distance, average passengers, and average market share are a point on the regression line.)
c. Using the uncertainty inherent in the data, perform a parametric Bootstrap to simulate the resulting average airfare. Compare the probability that airfares will rise (as a result of consolidation) for your models with and without parameter uncertainty.

5.3 PORTFOLIO CORRELATION

Exercise5-3.xlsx contains data for the weekly changes in the Shanghai Composite Index and the Dow Jones Industrial Average over the decade of 2000–2010.

a. Fit a distribution, without parameter uncertainty, for each index's weekly change and a parametric copula without uncertainty for the relationship between the two. Use the best AIC score to choose the distributions and copula fits.
b. Simulate the weekly results for each index over 4 weeks.
c. Simulate a $100 investment in the SSEC, a $100 investment in the DJIA, and a portfolio of $50 invested in each. Compare the probabilities of losing money for each investment.

5.4 INCOME AND HOUSING

Exercise5-4.xlsx has data from the 2000 census for median household income, median rent, and median owner occupied housing values by state.

a. Fit a Multivariate copula to the three variables using the best SIC score. Also, create a valid correlation matrix for the three variables.
b. Consider that a Bivariate copula fits for each pair of variables. Do the best fitting Bivariate copulas match the results in a?

5.5 TECHNOLOGY DIFFUSION

The Bass Diffusion model is commonly used to forecast adoption of new technologies. It rests on three assumptions: that there is a fixed population of users (who will eventually adopt the new technology), that *innovators* will adopt in proportion to the current number of adopters, and that there are *imitators* whose adoption is proportional to the product of the current adopters and the remaining potential adopters. The origins of the Bass Diffusion model lies in epidemiology: Infections may spread through a population in a similar fashion (to new technologies or services); innovators are people prone to getting a disease while imitators catch the disease from the innovators. The number of imitators over time is affected by two factors, working in opposite directions. There are more innovators to "infect" future imitators, but there are fewer potential adopters left to infect. Equation 5.1 reflects the Bass Diffusion model, with the assumption that the rate of innovation (p) and the rate of imitation (q) are constant over time:

$$n_t = p(N - N_{t-1}) + q\left(1 - \frac{N_{t-1}}{N}\right) \tag{5.1}$$

where n_t is the number of new adopters during period t, N_{t-1} is the cumulative adopters at the beginning of period t, and N is the total pool of potential adopters.

Equation 5.1 states that the new adopters will come from the innovators (a constant fraction of the remaining nonadopters) and the imitators (a constant proportion of the product of the current adopters and remaining fraction of nonadopters). Equation 5.1 is not in the best form to estimate from actual data, however. Multiple regression can be used to estimate the diffusion curve when it is rewritten in the form of Equation 5.2:

$$n_t = pN + (q - p)N_{t-1} - \left({q}/{N}\right)N_{t-1}^2 \tag{5.2}$$

Equation 5.2 only requires data on total adopters at each point in time, and using multiple regression analysis to estimate $Y = a + bX + cX^2$ will permit estimation of the parameters of the Bass Diffusion model.

a. Exercise5-5.xlsx provides annual data for the number of cellular subscribers in the United States over the 1988–2009 period. The total number of subscribers is given, along with its squared value, the number of new subscribers, the lag of the number of subscribers, and its squared value. Also shown is the average bill (but we will not use that in this problem). Estimate the Bass Diffusion model by fitting a multiple regression model to these data and then deriving p and q from the estimated coefficients of that model. Assume that the total number of potential adopters is 350,000,000.
b. Simulate the 2010 end-of-year total subscriber number, using the regression model and the uncertainty in the data. Provide a 95% confidence interval for your prediction.

TABLE 5.3

Fish Regeneration as a Function of Population Density

Density	Regeneration
0	0
0.1	50
0.2	100
0.3	200
0.4	320
0.5	500
0.6	550
0.7	480
0.8	300
0.9	180
1	0

5.6 FISH POPULATION

A simplified version for the biology of a fish population is to assume that the regeneration of fish is a nonlinear function of the fish density. Assume that Table 5.3 provides empirical data describing this relationship. Assume that the regeneration function is quadratic (a function of fish density and fish density squared), that the initial fish population is 500, and that the maximum population is 4,000 (used to determine the density at any point in time). Regeneration accounts for both births and natural deaths.

a. Build a regression model using the uncertainty inherent in the data to simulate the evolution of the fish population over 20 years. (*Hint:* add a column to your data for the density squared and then use a multiple regression with regeneration as the response variable and density and density squared as the independent variables.)
b. Assume that there are seven fishing ships targeting this species of fish and that each ship can initially catch 12 fish per year. Due to improved technology, this fish catch rate will increase by one each year (for each ship). Provide a population trend chart for 75 years, an estimate of the probability of extinction by year 50, and a histogram for the year of extinction for this population.

5.7 WARRANTIES REVISITED

Consider Exercise 4.7 in the preceding chapter concerning tire failures and warranties. Suppose that tread wearout and sidewall failures may be correlated. Use a Bivariate Normal copula to show how linear relationships with correlation ranging between zero and one (in increments of 0.1) affect your answer to Exercise 4.7 for the 100,000-mile warranty tire. Explain the results you observe.

5.8 THE LONG TAIL

The Long Tail refers to the idea that less popular items may, in the aggregate, offer as much potential sales as more popular items and that digital technologies make this "long tail" of the distribution available to suppliers.* Potential applications include movies, books, web searches, music, and even kitchen appliances. Excercise5-8.xlsx contains data on the top 100 movie DVD sales for 2009.

a. Related concepts include Power Laws and a special form known as Zipf's Law, which states that the frequency with which something occurs is proportional to one divided by the rank. Mathematically, this is equivalent to saying that the log (frequency) is a linear function of the log (rank). Investigate the DVD sales data (rank and number sold) for 2009 to see if they graphically look like they obey Zipf's Law. Estimate a linear regression model to test this.
b. The concept of the Long Tail relies on this apparent relationship between rank and number sold. Shelf space in physical stores may limit sales to the most popular 100 movie titles, but a digital store can advertise virtually every movie title. Use your estimated relationship between rank and number sold, and the uncertainty inherent in the data, to simulate the total number of sales that could result for ranks 101 through 1,000. Assuming that a digital store's sales are proportional to the entire market, estimate the proportion of total sales (by volume, not by dollars) for such a store that would come from the titles ranked 101 through 1,000.

* Anderson, C. 2006. *The Long Tail*. New York: Hyperion.

6

Time Series Models

LEARNING OBJECTIVES

- Appreciate the difference between modeling distributions versus time series.
- Understand time series forecasting as a data-driven process that is contingent on having a stable underlying pattern in the data.
- Recognize and capture trends, seasonality, and volatility in historical data.
- Learn how to fit time series models to data.
- Learn how to capture the uncertainty in a time series.
- Learn how to simulate time series data for future predictions.
- See how to use and incorporate judgment into time series models.
- Learn how to model several interrelated time series.

6.1 Introductory Case: September 11 and Air Travel

The September 11, 2001, terrorist attacks had pronounced effects on air travel. Among the many questions the incidents raised, two are particularly amenable to quantitative analysis and simulation:

- How much did the attacks affect air travel?
- Have the effects diminished over time?

Figure 6.1 shows monthly total revenue passenger mile (RPM) data for U.S. carriers from 1994 through 2006.*

This time series exhibits several typical features of time series: trend, seasonality, and randomness. The general upward trend is clear, as is, notably, the impact of the September 11 event. Seasonality is clear from the repeated annual cyclical pattern in the data. Randomness means that the patterns in

* Data from the U.S. Department of Transportation, Research and Innovation Technology Administration, Bureau of Transportation Statistics, http://www.transtats.bts.gov/ (accessed on December 31, 2010).

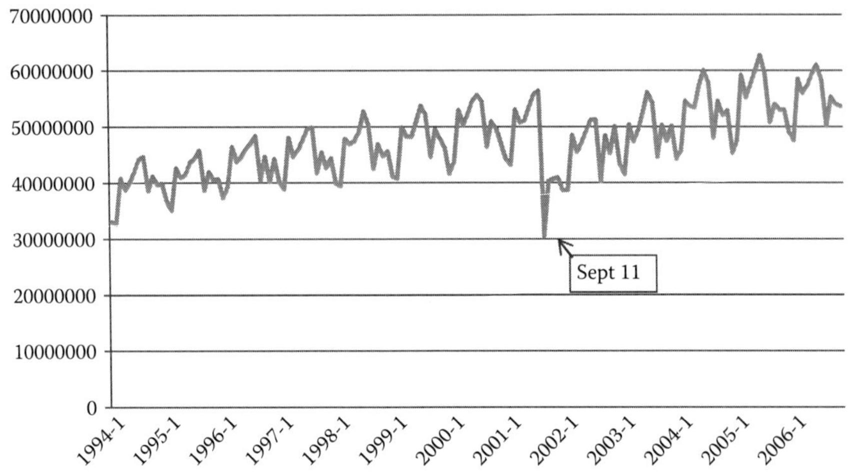

FIGURE 6.1
U.S. air passenger travel.

the time series are not perfectly repeated. Appropriate analysis of these data can help estimate the impact of the September 11 event, as well as indicate whether the effects of the event have been wearing off over time. Before we examine these questions, we must address how time series analysis differs from the other types of analyses we have discussed thus far.

6.2 The Need for Time Series Analysis: A Tale of Two Series

Imagine we are collecting data where we measure the value of a variable once every interval (day, week, year, etc.) over a period of time. Figure 6.2 shows a histogram (the bars indicating the frequency of values) of the hypothetical data, and Figure 6.3 shows the same set of values as they were collected chronologically through time.

Figure 6.2 shows that when a distribution is fitted to them, the data are approximately normally distributed, and a 95% confidence interval is approximately (80,160). The mean (120) and the confidence interval are superimposed on the time series in Figure 6.3. Suppose that we wanted to use these data as a basis for a simulation, but we did not know (or did not consider) that the data were taken as a sequential series of values. In that case, we might only fit a distribution to the data (Figure 6.2) and ignore the temporal (i.e., time related) nature of the data. The result of this would be to underestimate the next predicted value(s) in the data series, and, if we ran an entire simulation, our confidence interval would be far too wide. Clearly,

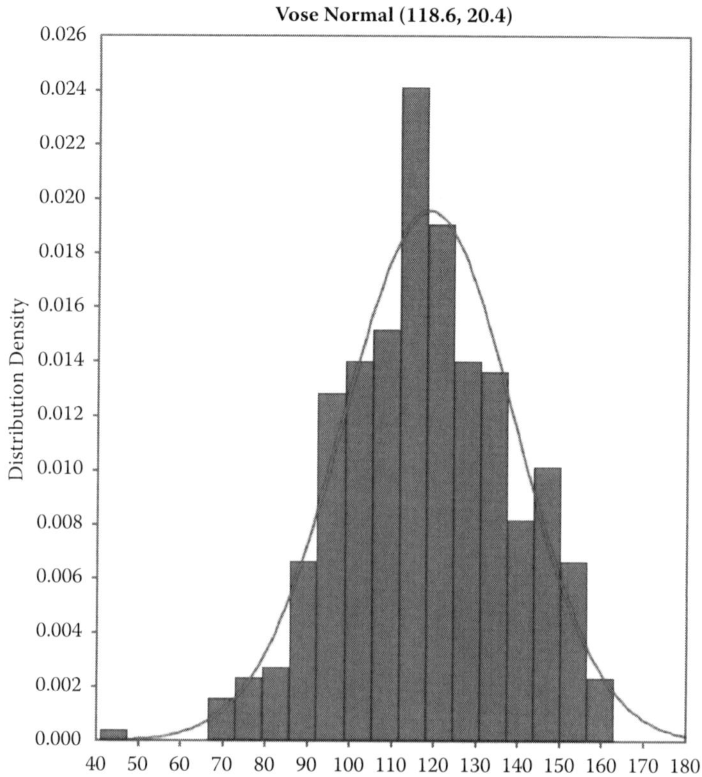

FIGURE 6.2
Histogram for hypothetical data series I.

there is a time pattern to these data, and failure to account for it would result in an inaccurate prediction that is far too uncertain. While this seems very logical within this simple problem, ignoring the temporal pattern when analyzing and forecasting based on the data is a very common mistake in Monte Carlo simulation.

Next, consider Figure 6.4, which shows a different hypothetical data set. Suppose that the points between the vertical lines indicate a period of time series data used to fit a distribution and then predict future values. We would predict a value of around 240 with a high degree of certainty (a 95% confidence interval from 225 to 255), as shown in Figure 6.5.

However, the full data set shows that our prediction for the next value (after the second vertical line) may well be too high (by a factor of two) and far more certain than it should be. Again, there is a time structure to the data that must be included in the analysis if we are to make reliable predictions.

In both of the preceding examples, the use of fitted distributions of time series data to make predictions results in poor outcomes, unless the temporal

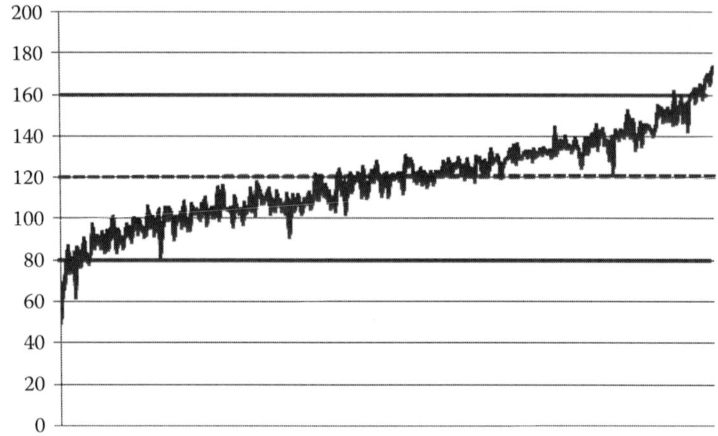

FIGURE 6.3
Time series for hypothetical data series I.

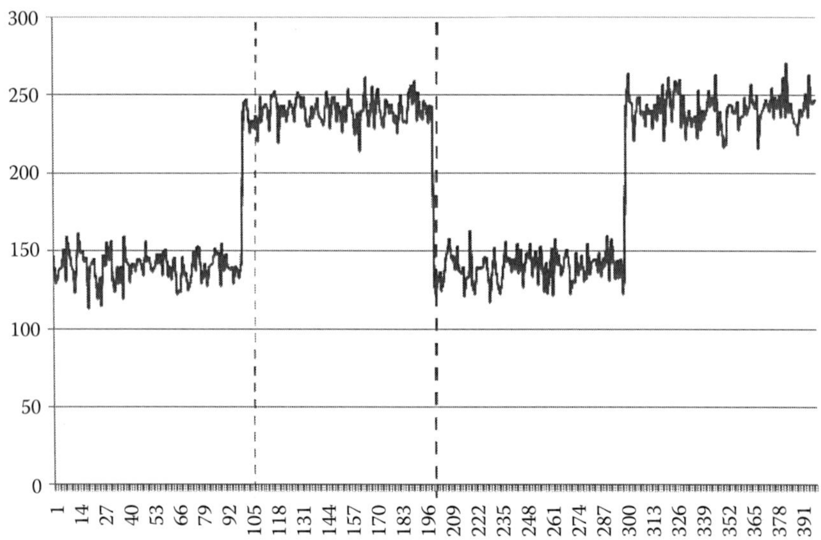

FIGURE 6.4
Time series for hypothetical data series II.

patterns are correctly modeled. The general approach to modeling data such as these is to try to separate the fundamental elements of the time series (e.g., level, trend, seasonality, randomness, and possibly even jumps or changes in volatility), simulate each element individually, and then recombine them to produce the final forecasting model.*

* Sometimes, cyclical factors can also be identified. These would refer to repeating patterns that do not follow traditional time units (e.g., days, weeks, months, etc.).

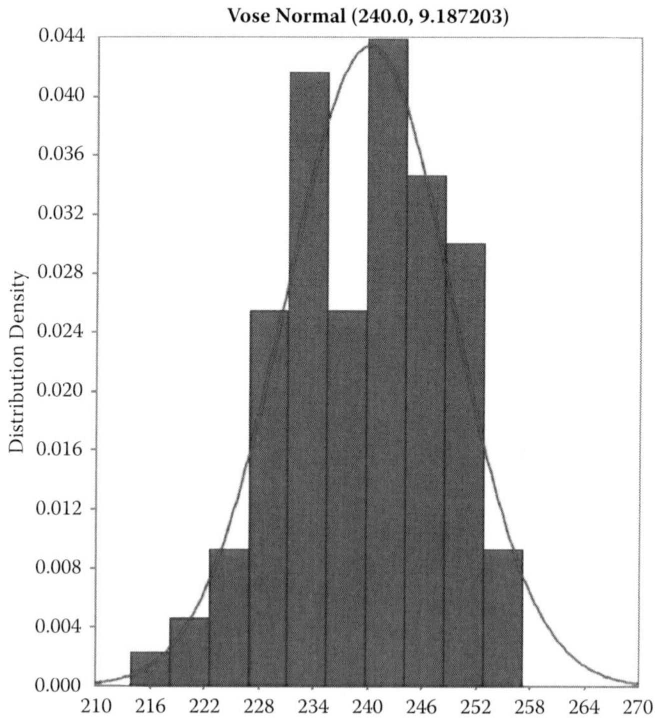

FIGURE 6.5
Forecast for hypothetical data series II.

6.3 Analyzing the Air Traffic Data

Our approach to quantifying the effect of the September 11 attacks is first to consider the time series data through August 2001, using those data to forecast the passenger traffic through 2006 that we would have expected if there had been no unusual event, and then compare the predicted air traffic volumes with the actual September 2001–December 2006 data.* The first step in time series analysis should always be to view the data (Figure 6.1). (You should always view your historical data and the forecast as well; see Figure 6.12.) To fit a time series model in ModelRisk, we use the Fit button on the ModelRisk toolbar and select Univariate Time Series; however, before we do this, we must first remove the obvious seasonality and trend in our data.†

* We generally do not recommend forecasting future values that are more than 30% of the available historical data—in this case, around 2 years. Since we have the data through 2006, we will extend our forecast that far, but with the caveat that forecasting this far beyond the historical data is typically not advised.

† Since there are no time series methods within ModelRisk to take into account seasonality, we need to do this "manually," as shown in this section.

Air travel6.xlsx contains the data and our models. Our first step is to construct a pivot table for the number of passengers by month and by year, as shown in Figure 6.6. The bottom table (cells H18:T32) calculates the monthly seasonal factors by dividing the actual monthly passenger traffic (e.g., cell I3 for January 1994) by the average traffic for that year (e.g., cell U3). The averages shown on row 32 only average the seasonal factors for the pre-September 11 time period (that is, before September 2001). Similarly, column V calculates the year-over-year percentage growth in annual traffic, with the average (cell V16) only using the pre-September 11 time period.

We will use these average seasonal and trend factors to account for (and remove) seasonality and trend from our data. More sophisticated approaches could certainly be used at this stage; for example, those that take into account the uncertainty we have about the seasonal or trend factors. Since our focus is on how to construct a time series forecast, we will simply use the averages of the seasonal and trend factors over the pre-September 11 period.

To remove seasonality from the data, the seasonal factors are used to inflate or deflate the historical data according to the month. Figure 6.7 shows the first several rows of this adjustment process. Column C contains the historical data, and the formula in column D removes the seasonality according to whether the seasonal factor indicates that the month (column B) is typically a month with below average or above average traffic. The HLOOKUP function looks up the month number in the pivot table in Figure 6.6 and reports the average seasonal factor from row 32 for that month. Column E uses the average annual percentage increase (cell V16) to remove the trend. The deseasonalized and detrended data for the pre-September 11 time period are shown in Figure 6.8.

Now we are ready to use ModelRisk to fit a time series model to the remaining variability in the time series. Using the data in cells E2:E93, using "Fit" "Univariate Time Series Models," and selecting all the possible models give the Fit window shown in Figure 6.9. As you can see in the figure, the MA1 time series has the highest –SIC, –AIC, and –HQIC. As discussed in Chapter 4, all three parameters are "Information Criteria," and they can best be seen as statistics that can help you determine which of the fitted time series models best fits your data.* They should not be used as the sole criterion of which model (or distribution, when fitting distributions) to use.

Based on the Information Criteria, the first order moving average time series model (MA1) provides the best fit and this is what we will use for this example.† Notice that several of the models cannot be fit to these data and error messages are provided for this in the lower left of the fitting window. There is an option for the "Number of lines," which is set by default to 1 when the window first opens. We have changed it to 2 and pressed the Generate button to show two potential forecasts. (We have changed the number of

---

* The Information Criteria are some of the many "goodness of fit" statistics.

† It is generally advisable to select only models that appear to be relevant for your data.

	H	I	J	K	L	M	N	O	P	Q	R	S	T	U	V
1	**Average of Actua**	**Column La**													
2	**Row Labels**	**1**	**2**	**3**	**4**	**5**	**6**	**7**	**8**	**9**	**10**	**11**	**12**	**Grand Total**	trend
3	1994	33063218	32706163	40829245	38672069	40160849	41916076	44225966	44685652	38567585	41256310	39649572	39868772	39633289.75	
4	1995	36640667	35092555	42646196	40973075	41441924	43628058	44438865	45788785	38624189	41986217	40507651	40698590	41038897.67	0.034851
5	1996	37303173	39187394	46464431	43770812	44680604	45996703	47059320	48379108	40401215	44666103	40168952	44288217	43535502.67	0.059056
6	1997	40094443	38873954	48029975	44674588	45801428	47672314	49646138	49859398	41701653	45443234	42634396	44459002	44907543.58	0.031029
7	1998	39907754	39393370	47886859	46917865	47408613	49157456	52776459	50430673	42459397	46921944	44751747	45629996	46136844.42	0.027006
8	1999	41036190	40719445	49893855	48297891	48166998	50899806	53705361	52182372	44613266	49501818	47908281	46288717	47767650	0.034737
9	2000	41557193	43729534	52990192	50354369	52325210	54724492	55621547	54515661	46398645	50958213	49659124	47075544	49992477	0.045524
10	2001	44109939	43180235	53058085	50794947	51122786	53473441	55605088	56405712	30546484	40290718	40691635	40901001	46698339.25	-0.06816
11	2002	38557639	38644502	48500814	45437855	47127122	49277700	51256869	51315219	40275540	48406823	45209349	50037190	46170551.83	-0.01137
12	2003	43352308	41468358	50390457	47368360	49415544	52544040	56147350	54324485	44578644	50351804	47459241	50134424	48961249.58	0.058887
13	2004	44209783	45712143	54627608	53714513	53393384	57352668	60067665	57792551	48001640	54582656	52059824	52911735	52868847.5	0.076785
14	2005	45237238	47360704	59199365	55228855	57594568	60014028	62726405	59408541	50854026	54022108	53014927	53011799	54806047	0.035986
15	2006	49044534	47525840	58523792	56014579	57312853	59440627	81009560	58481309	50154832	55299297	54059204	53684316	55045894.42	0.004367
16	**Grand Total**	**41085698.38**	**41045707.46**	**50233913.38**	**47863059.85**	**48919375.62**	**51238262.23**	**53422044.85**	**52582265.08**	**42864393.54**	**47975926.54**	**45982607.92**	**46845023.31**	**47504856.51**	0.038701
17															
18	Seasonal Factors	1	2	3	4	5	6	7	8	9	10	11	12		
19	1994	0.834228453	0.825219486	1.030175523	0.975747137	1.013311013	1.057597698	1.115879259	1.127477741	0.973110868	1.040950934	1.000410823	1.005891064		
20	1995	0.892827758	0.855104718	1.039165241	0.998396091	1.009820594	1.063090397	1.082847433	1.115741104	0.941160489	1.02308345	0.987055045	0.99170768		
21	1996	0.85684489	0.900124992	1.067276778	1.005404999	1.026302702	1.056533178	1.080941235	1.111256447	0.929384353	1.025969617	0.922671143	1.017289667		
22	1997	0.892822003	0.865644186	1.069530221	0.994812551	1.019904995	1.061565835	1.105518896	1.110267764	0.928611313	1.01192874	0.949381609	0.990011888		
23	1998	0.864986639	0.853837546	1.037930955	1.016928349	1.02758514	1.065470702	1.143911329	1.093067236	0.920292611	1.017016759	0.969978497	0.989014237		
24	1999	0.859079105	0.852448153	1.0445114	1.01110042	1.008360219	1.065570653	1.124304022	1.092420749	0.933964011	1.036300048	1.002944064	0.968997156		
25	2000	0.831268933	0.874722291	1.059963322	1.007238929	1.046661681	1.094654542	1.112598342	1.090477293	0.928112544	1.019317627	0.993331937	0.941652561		
26	2001	0.944571899	0.924663183	1.136187836	1.087724913	1.09474527	1.145082285	1.195012262	1.207874047	0.654123562	0.862786957	0.871372208	0.87585558		
27	2002	0.835113237	0.836994588	1.050470745	0.984130646	1.020718188	1.067297185	1.110163664	1.111427457	0.872320958	1.048435011	0.979181474	1.083746848		
28	2003	0.885441208	0.84696282	1.029190583	0.967466321	1.009278652	1.073176041	1.146771181	1.109540003	0.910488282	1.028401122	0.969322503	1.023961284		
29	2004	0.836216129	0.864632863	1.033260481	1.015995535	1.009921467	1.084810256	1.136163693	1.093130525	0.907938082	1.032416226	0.984697539	1.000811206		
30	2005	0.825405963	0.864151067	1.080161191	1.007714023	1.050879805	1.095025664	1.144516133	1.08397785	0.927890785	0.985696122	0.967318935	0.967261861		
31	2006	0.890975331	0.863385735	1.06318178	1.017597763	1.041183064	1.079837609	1.108339698	1.062410006	0.911145736	1.004603478	0.982075132	0.975264669		
32	averages	0.87207871	0.868970569	1.06059266	1.012169174	1.030833952	1.076195661	1.120126597	1.118572798	0.936376598	1.024938168	0.975110445	0.986366322		

FIGURE 6.6
Pivot table.

	A	B	C	D	E
1	year	month	Actual RPM	seasonally adjusted	detrended
2	1994	1	33063218	=C2*(1/HLOOKUP(B2,$I$18:$T$32,15,FALSE))	=D2*(1/((1+$V$16)^(A2-1994)))
3	1994	2	32706163	=C3*(1/HLOOKUP(B3,$I$18:$T$32,15,FALSE))	=D3*(1/((1+$V$16)^(A3-1994)))
4	1994	3	40829245	=C4*(1/HLOOKUP(B4,$I$18:$T$32,15,FALSE))	=D4*(1/((1+$V$16)^(A4-1994)))
5	1994	4	38672069	=C5*(1/HLOOKUP(B5,$I$18:$T$32,15,FALSE))	=D5*(1/((1+$V$16)^(A5-1994)))
6	1994	5	40160849	=C6*(1/HLOOKUP(B6,$I$18:$T$32,15,FALSE))	=D6*(1/((1+$V$16)^(A6-1994)))
7	1994	6	41916076	=C7*(1/HLOOKUP(B7,$I$18:$T$32,15,FALSE))	=D7*(1/((1+$V$16)^(A7-1994)))
8	1994	7	44225966	=C8*(1/HLOOKUP(B8,$I$18:$T$32,15,FALSE))	=D8*(1/((1+$V$16)^(A8-1994)))
9	1994	8	44685652	=C9*(1/HLOOKUP(B9,$I$18:$T$32,15,FALSE))	=D9*(1/((1+$V$16)^(A9-1994)))
10	1994	9	38567585	=C10*(1/HLOOKUP(B10,$I$18:$T$32,15,FALSE))	=D10*(1/((1+$V$16)^(A10-1994)))
11	1994	10	41256310	=C11*(1/HLOOKUP(B11,$I$18:$T$32,15,FALSE))	=D11*(1/((1+$V$16)^(A11-1994)))
12	1994	11	39649572	=C12*(1/HLOOKUP(B12,$I$18:$T$32,15,FALSE))	=D12*(1/((1+$V$16)^(A12-1994)))
13	1994	12	39866772	=C13*(1/HLOOKUP(B13,$I$18:$T$32,15,FALSE))	=D13*(1/((1+$V$16)^(A13-1994)))
14	1995	1	36640667	=C14*(1/HLOOKUP(B14,$I$18:$T$32,15,FALSE))	=D14*(1/((1+$V$16)^(A14-1994)))
15	1995	2	35092555	=C15*(1/HLOOKUP(B15,$I$18:$T$32,15,FALSE))	=D15*(1/((1+$V$16)^(A15-1994)))
16	1995	3	42646196	=C16*(1/HLOOKUP(B16,$I$18:$T$32,15,FALSE))	=D16*(1/((1+$V$16)^(A16-1994)))
17	1995	4	40973075	=C17*(1/HLOOKUP(B17,$I$18:$T$32,15,FALSE))	=D17*(1/((1+$V$16)^(A17-1994)))
18	1995	5	41441924	=C18*(1/HLOOKUP(B18,$I$18:$T$32,15,FALSE))	=D18*(1/((1+$V$16)^(A18-1994)))
19	1995	6	43628058	=C19*(1/HLOOKUP(B19,$I$18:$T$32,15,FALSE))	=D19*(1/((1+$V$16)^(A19-1994)))
20	1995	7	44438865	=C20*(1/HLOOKUP(B20,$I$18:$T$32,15,FALSE))	=D20*(1/((1+$V$16)^(A20-1994)))
21	1995	8	45788785	=C21*(1/HLOOKUP(B21,$I$18:$T$32,15,FALSE))	=D21*(1/((1+$V$16)^(A21-1994)))
22	1995	9	38624189	=C22*(1/HLOOKUP(B22,$I$18:$T$32,15,FALSE))	=D22*(1/((1+$V$16)^(A22-1994)))
23	1995	10	41986217	=C23*(1/HLOOKUP(B23,$I$18:$T$32,15,FALSE))	=D23*(1/((1+$V$16)^(A23-1994)))
24	1995	11	40507651	=C24*(1/HLOOKUP(B24,$I$18:$T$32,15,FALSE))	=D24*(1/((1+$V$16)^(A24-1994)))
25	1995	12	40698590	=C25*(1/HLOOKUP(B25,$I$18:$T$32,15,FALSE))	=D25*(1/((1+$V$16)^(A25-1994)))

FIGURE 6.7
Removing seasonality and trend.

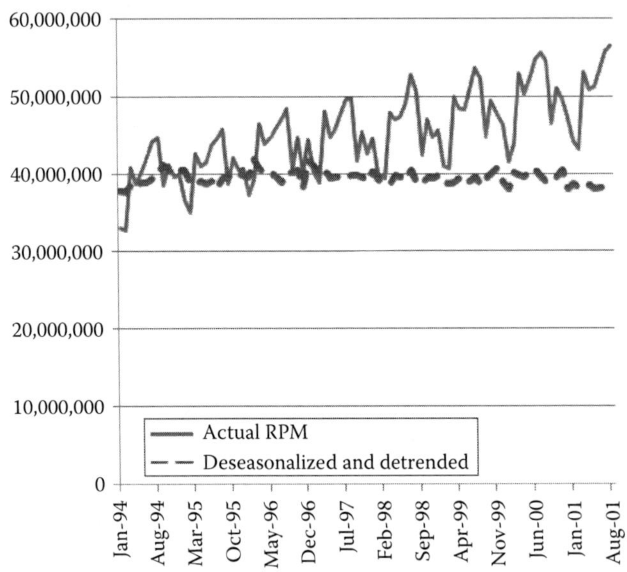

FIGURE 6.8
Deseasonalized and detrended data.

forecast periods to equal 64; it should be set to match the time horizon of interest.) Showing multiple forecasts gives an initial idea of how much variability there will be in the fitted time series model. We have also selected the "Fix Y-scale" radio button, which facilitates comparison of different forecasts by preventing the scale from changing each time you press the Generate

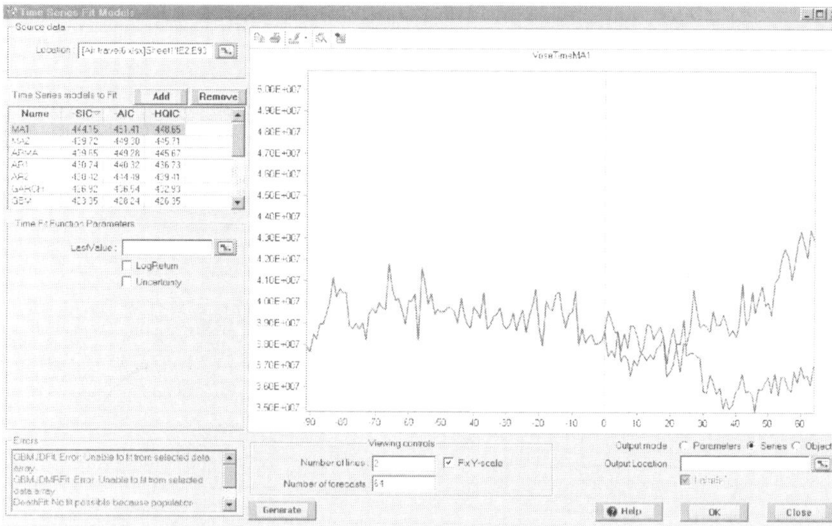

FIGURE 6.9
Fit time series dialog box.

button. As with fitting distributions, there is an option to include Uncertainty in the forecast model (not selected here). Also, as with fitting distributions, it is important to inspect the best fitting model visually rather than to rely only on the Information Criteria to rank the models.

There are a number of options for pasting the forecasts into the spreadsheet. Since we are interested in the 64 forecasts, we can paste the "Series" into a range of 64 spreadsheet cells (pasted as an array function). We will use the alternative of pasting an "Object" into the spreadsheet—but using the LastValue button (on the left side of Figure 6.9) by using cell E93, as shown in Figure 6.10.

This will permit us to forecast the post-September 11 traffic. We use the VoseTimeSimulate function to simulate the data (after January 1994) based on this Object (placed in cell H37). It is important that VoseTimeSimulate be used, rather than VoseSimulate, and that it be entered as an Excel array function. This ensures that the autocorrelation over time of the forecasts is correctly modeled. The simulation formulas are shown in Figure 6.11, along with the formulas to recombine these simulated values with the seasonal and trend factors.

Column F produces the simulated monthly traffic from the time series Object in cell H37; the curly brackets ({ }) in the formula bar show that it was entered as an array formula. Column G adds the seasonal factor and trend (based on the month in column B and the year in column A) to these simulated values to produce the simulated traffic for the post-September 11 time period. Figure 6.12 shows a single random simulation of the time series.

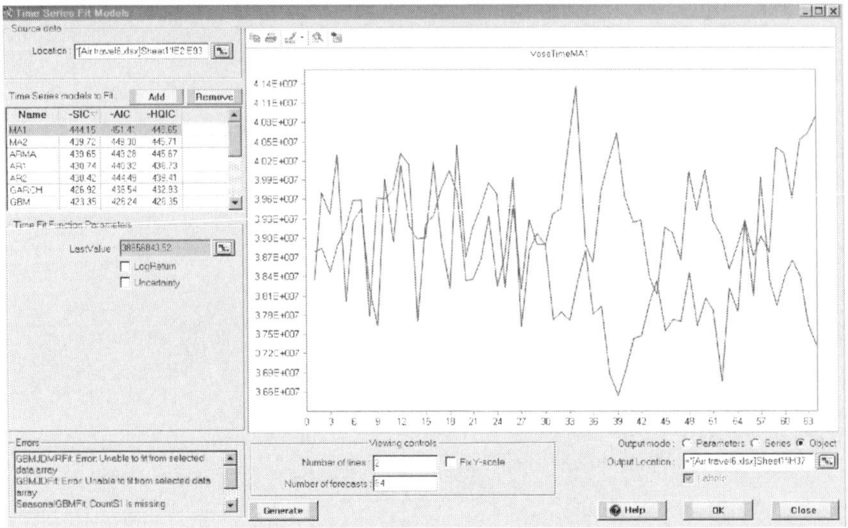

FIGURE 6.10
Fitting an Object with a last known value.

To answer our original questions about the impact of September 11 on air passenger traffic, we calculate the shortfall for September 2001–December 2006 by subtracting the actual traffic (column C) from the simulated traffic (column G). We place this in cells H94:H157 and designate that range as a ModelRisk Output range. Running the simulation produces the trend chart shown in Figure 6.13. There are two striking findings in this figure. First, median shortfall is generally around 4 million to 8 million revenue passenger miles per month, a significant impact (on the order of 10%). Second, the shortfall does not appear to decrease after the first couple of months. Apparently, September 11 caused more than a temporary aberration in air traffic.

There are two important caveats to this conclusion. First, we have not investigated any other possible causes for a decline in air traffic, such as changes in corporate budgets devoted to air travel (and the potential rise in alternatives to physical air travel using information technology, such as video conferencing). Second, we have used 8 years of data to predict an additional 5 years—an excessive forecast period for the available data. The implicit assumption is that the historical patterns of seasonality, trend, and variability would remain unchanged over this period.

It is also useful to have a measure of the degree to which our time series model provides a good description of the true traffic. *Thiel's U* is a statistic that can provide such a measure. This measure is based on comparing the output of a stochastic time series model to the time series of actual data. Thiel's U calculates the total sum of the squared deviations between the forecasted time series and the actual data. The closer Thiel's U is to zero, the better the

FIGURE 6.11
Simulating variability and adding trend and seasonality.

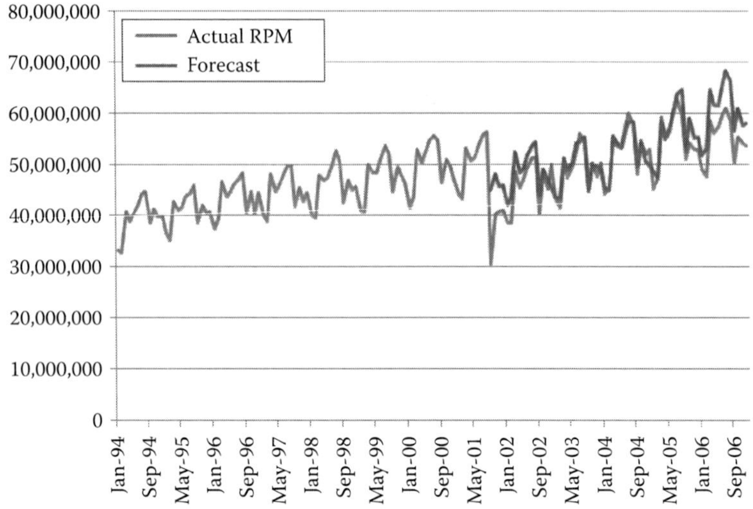

FIGURE 6.12
Simulated air traffic.

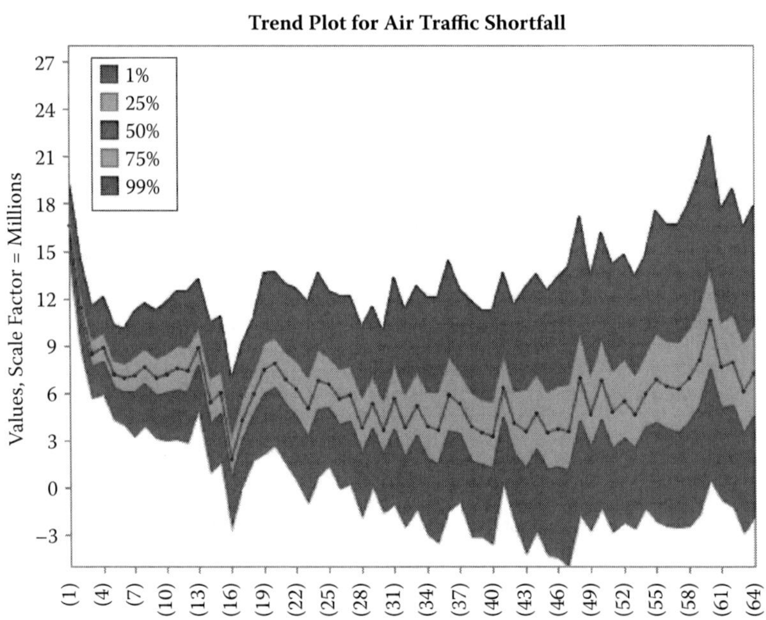

FIGURE 6.13
Trend chart for forecasts.

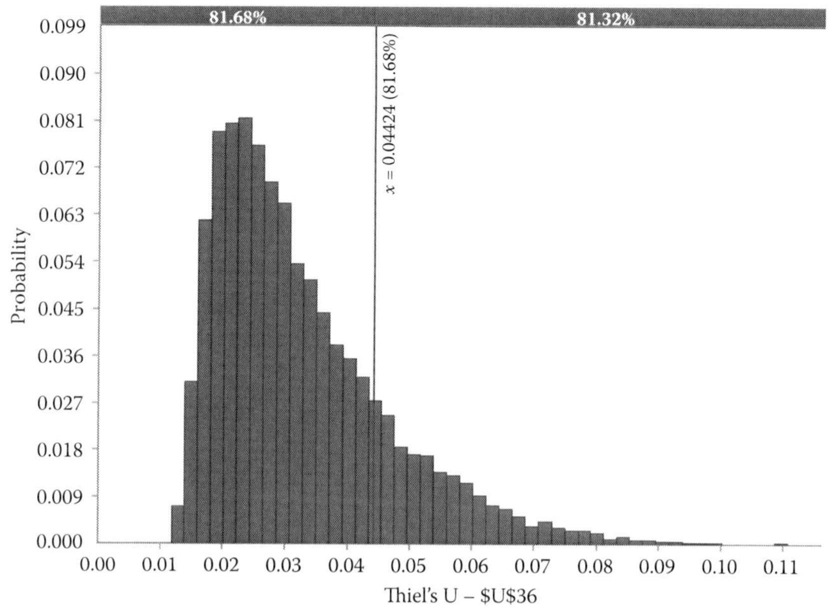

FIGURE 6.14
Thiel's U.

forecast is. When Thiel's U equals one, the chosen model is no better than a naive forecasting model, such as simply repeating the last observed value.

To implement Thiel's U in our air travel model, we calculate it for pre-September 11 forecasts ("backcasts") and a naive forecast, simply based on projecting each month's traffic volume as the predicted traffic for the next month. We use the VoseThielU function for the naive forecast and get the value .044235. For our stochastic forecast, the value will vary, so we mark it as a ModelRisk output. Figure 6.14 shows the results: In almost 82% of the iterations, our forecasting model outperforms the naive forecasting model, thus providing us with a fairly high level of confidence that our predictive model is better than a naive forecast.

6.4 Second Example: Stock Prices

Forecasting stock prices is of considerable interest for many people, particularly those with equity investments. Stock prices6.xlsx contains 10 years of daily stock price data for AT&T (ticker symbol: T). Figure 6.15 shows the daily

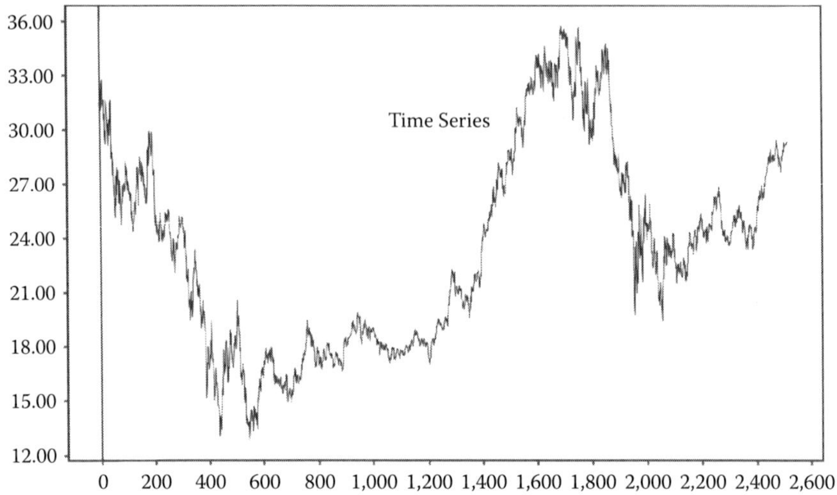

FIGURE 6.15
AT&T stock price data.

closing price and Figure 6.16 shows the percentage *change* in AT&T's stock price, calculated as the difference in natural logarithms of successive daily prices.*

The pattern in Figure 6.16 is typical for daily changes in a stock price. The changes seem to fluctuate randomly around a mean close to zero. If there were clear trends that could be counted on continuing into the future, we would expect investors to have discovered them, acted on that knowledge (i.e., bought or sold the stock), and, as a result, destroyed the trend. As Figure 6.16 shows, most of the time, the daily fluctuations are within a range of ±3%, although there are occasional changes on the order of 6%. We will now compare fitting a *distribution* to the stock price *change* values with two alternative time series models. As we will see, the results will be fairly similar, but not quite the same.

Figure 6.17 shows the results of fitting time series models to these data. It was necessary to check the LogReturn box so that ModelRisk knows that these are log changes (and that zero and negative values are to be expected). The best fitting model is the GARCH model.† We also fit a moving average model (MA2) and a static Normal distribution to the data values (the latter ignoring the time structure). The comparisons between these forecasts will be instructive. We simulate 100 days of prices and define ModelRisk output ranges for the forecasts. Running the simulation and comparing the first day's price forecast for the three models gives Figure 6.18.

* This is calculated as (ln(pricedayt) – ln(priceday(t–1), which equals ln(pricedayt/priceday(t–1), which is almost the same as the discrete calculation of a percentage change—(pricedayt – priceday(t–1))/priceday(t–1)—for small changes.

† GARCH stands for *generalized autoregressive conditional heteroskedasticity time series model.*

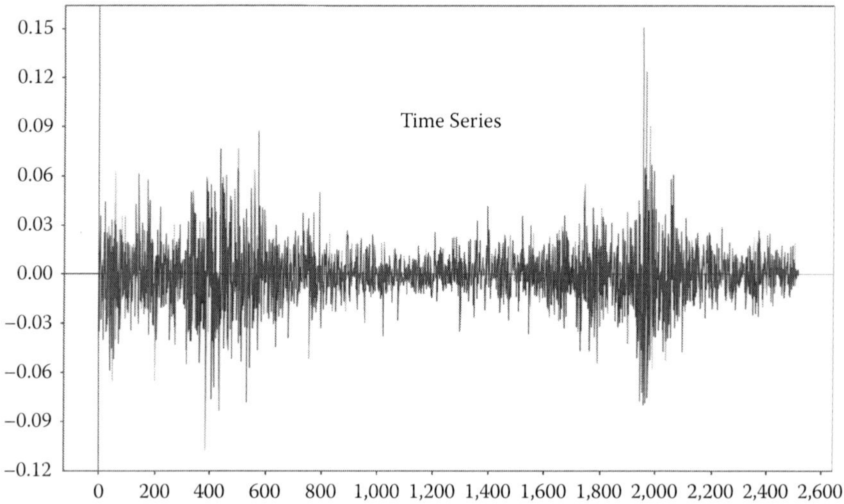

FIGURE 6.16
Changes in AT&T stock prices.

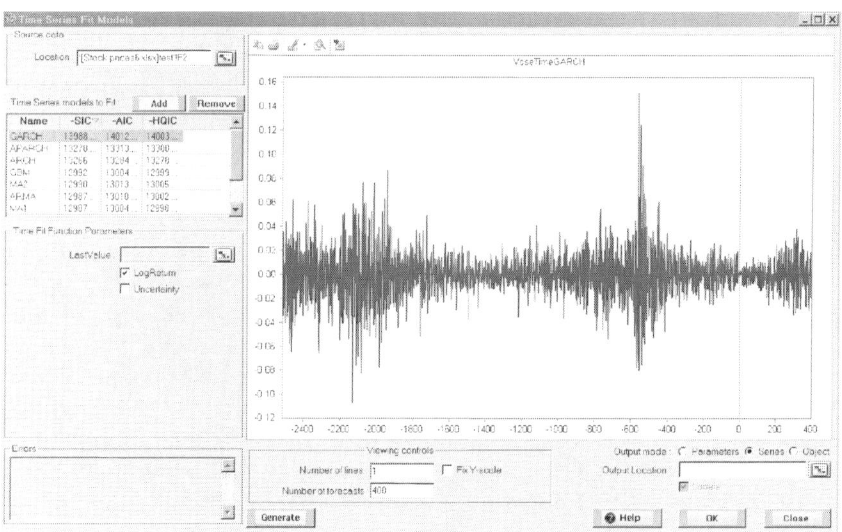

FIGURE 6.17
Stock price time series models.

Note that the moving average and static normal distribution show almost equivalent results that extend throughout the forecast period. In this sense, there is little gained by using time series methods for stock price changes rather than just fitting a static distribution to the values. However, the GARCH model provides a narrower forecast. This is easily seen in Figures 6.19 and

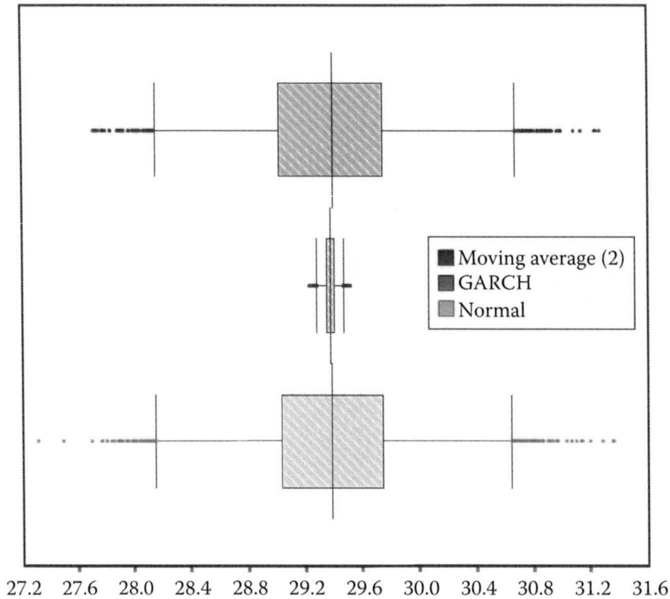

FIGURE 6.18
Comparison of three stock price time series models.

6.20, which show the trend charts* for the 100 forecast day prices using the GARCH and MA2 models, respectively.

The moving average model predicts prices in the range of \$20–\$43, while the GARCH model exhibits a much narrower range (\$25–\$35). The reason that the GARCH model produces a smaller range of forecast prices is because it picks up the fact that the recent volatility in the stock price has been relatively low. The GARCH model is designed to model changes in volatility over time, and if you think this is an accurate reflection of the stock price behavior, it would be a good forecasting model to use. Both models correctly show increasing uncertainty over the forecast period. Figure 6.16 clearly shows that the AT&T stock price goes through periods of relatively high volatility interspersed with periods of low volatility. The moving average model cannot capture this and neither can a fitted probability distribution. So, while the level of a stock price is highly uncertain, there are time series methods

* To enable trend charts in ModelRisk, highlight the range of cells and designate them as model outputs, giving them a range name. After the simulation is run, you can open the forecast for any individual output cell in the series or display the entire series as a trend chart (by choosing this option in the results window). Figures 6.19 and 6.20 were constructed manually so that the scales could be made the same between them. ModelRisk trend charts automatically scale to the data. (The two time-series models differ considerably in the range of the forecasts.)

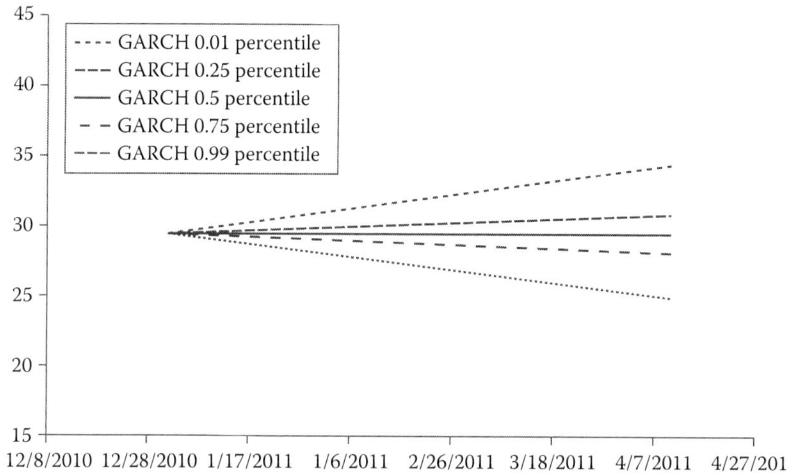

FIGURE 6.19
Trend chart for the GARCH model.

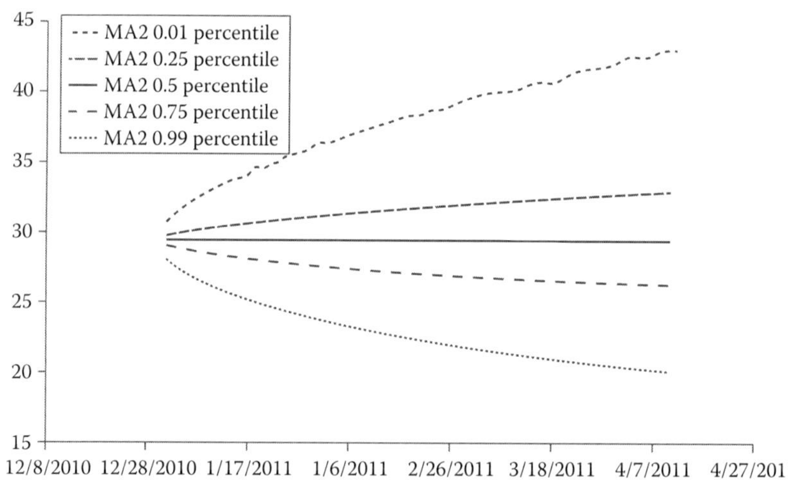

FIGURE 6.20
Trend chart for the MA2 model.

that can take into account changes in the degree of volatility. Whether the GARCH or MA2 model is "better," however, is a different matter. In this case, the GARCH fits the data better (according to the Information Criteria), but the ultimate test of which model is better is which does a better job of predicting the true stock price.

6.5 Types of Time Series Models

A wide variety of stochastic time series models have been developed, but most can be grouped into a few major types.

Geometric Brownian Motion (GBM) models. These models assume that the natural log of the *changes* (i.e., log returns) in a variable between consecutive periods are independent and follow a normal distribution. These GBM models are sometimes referred to as "random walk models" since each step in the time series is calculated as a random change from the prior value. The change that occurs in one period is not affected by, and does not affect, any previous or future changes. Also, the Black-Scholes formula to calculate option values assumes that the price series follows a GBM model.

Autoregressive Moving Average (ARMA) models. These assume that the expected (natural log) change in a variable is proportional to its change in the previous recent periods (autoregression) and that the expected change of the variable differs from the long-run mean by a factor proportional to its recent variation (moving average) from its long-run mean. In other words, within these time series methods, the change that occurs in one time period is affected by what happened in the last one or two periods.*

Autoregressive Conditional Heteroskedasticity (ARCH) models. Within these time series, the volatility of the time series is a function of the previous deviations of the variable from its long-run mean. In other words, in these time series the volatility can change over time.

Each general class of time series model has several subset models, generally based on whether one-period or two-period dependence in the series is assumed. Some accommodate seasonal patterns, while others permit large random jumps in the series—*jump diffusion.*

Expert Opinion-Based models. In practice, most time-series forecasts are not purely based on historical data, but rather on expert opinion or a combination of historical data and expert opinion. There are numerous time series methods and models that can be used to simulate forecasts based on expert opinion. ModelRisk contains a few tools that can be used for purely judgmental time series modeling based on expert opinion. These are found under the Time Series button on the ModelRisk toolbar by selecting SME Time Series. There are five variations available within ModelRisk:

- Poisson: for modeling events that occur randomly over time
- 2Perc: for modeling where there are estimates for an upper and lower percentile

* ARMA models generally can take into account what happened more than two steps ago, but within ModelRisk they are limited to looking back only two steps.

- Three Point: for modeling where there are minimum, most likely, and maximum value estimates over time (a time series version of the PERT distribution; there is also an option to use the Triangle distribution)
- Uniform: for modeling when there are minimum and maximum estimated values only
- Saturation: for modeling the number that will possess a particular characteristic (e.g., buy a product, catch a disease, etc.) from a fixed population base

These SME time series methods all permit autocorrelation over time to be specified by a single parameter. They provide automated methods to use expert opinion to develop time series simulations, without requiring the choice of a particular time series model (such as GARCH).

There are also two ModelRisk functions, VoseTSEmpiricalFit and VoseTimeEmpiricalFit, that fit time series to data, making few assumptions about the underlying process creating those values.* As stressed in Chapters 4 and 5, closely replicating past patterns is only desirable if you expect the future to resemble the past; otherwise, more judgmental modeling is required.

To elaborate on this point, the time series models we have illustrated so far (the expert time series being an exception) are all based on historical data.† That is, there is fairly limited judgment involved in the modeling.‡ The assumption is that future patterns and time series behavior will be similar to those in the past and that the available historical data contain the information necessary for making future forecasts. If you believe that the past patterns in the data under consideration may be misleading or are not likely to be replicated in the future, then you should not rely too heavily on these forecasting models. It is possible, however, to use past history as a guide to future patterns and to then utilize other additional assumptions to modify the forecast for the future. The next section provides an example of this hybrid approach.

* These functions are nonparametric. The only assumption they make about the underlying dynamic process is that it is "memoryless": The value at any point of time is independent of the history that preceded it.

† One other frequently used technique is to smooth a data series using *exponential smoothing*. This method assumes that each forecast in the time series is a weighted average of the current observation and past observations. The smoothing constant defines how much weight is placed on the recent past compared with the more distant past. However, choice of the smoothing constant requires the modeler's judgment or can be estimated with a regression model (to choose the smoothing constant that minimizes the sum of the squared deviations of the data series from the forecasts). In contrast, the stochastic time series models we discuss in this chapter use maximum likelihood methods to best fit a forecast model to the actual observations.

‡ There certainly is judgment involved in forecasting, including what historical data period to use for fitting a time series model, which exact time series method to use, and what patterns to take into account.

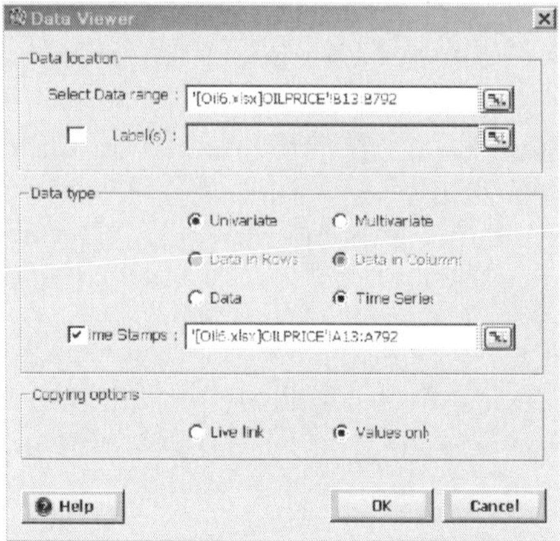

FIGURE 6.21
The data viewer.

6.6 Third Example: Oil Prices

Oil6a.xlsx contains monthly oil prices from 1946 through 2010. To get an initial view of the data, we could use the Data Viewer on the ModelRisk toolbar, as shown in Figure 6.21. Choosing the Univariate Time Series tab in the resulting window provides Figure 6.22. Of particular note is the bottom panel showing the moving standard deviation of this time series. Time series forecasting methods assume that the underlying historic process is stable; it will fluctuate randomly, may exhibit a trend and/or seasonality, and may periodically have random jumps, but there is some overarching stable process that can include these features.*

Oil prices, in contrast, appear to have undergone some fundamental changes during this 64-year period, and this is corroborated by important changes in market conditions. This time period is too long for good analysis because the Organization of Petroleum Exporting Countries (OPEC) pricing policies really began in 1974, which is where we see price level and volatility begin to increase. Prices have been volatile, but it might be reasonable to assume that the underlying process has been relatively stable since then—governed by oil supplies, OPEC behavior, drilling and extraction technologies, and worldwide demand (particularly in emerging economies). Indeed, if we confine our view to the 1974–2010 time periods, we get Figure 6.23.

The most recent period still exhibits a marked increase in volatility, probably indicating changes in the underlying market conditions. Oil companies

* For example, GARCH models can account for changes in volatility, although the assumption is that there is a stable pattern that underlies these changes in volatility.

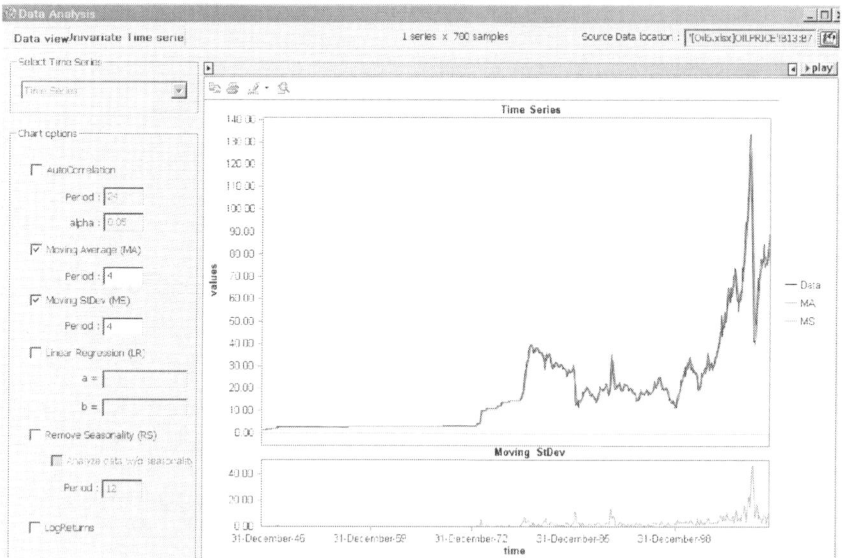

FIGURE 6.22
Oil prices data view.

make decisions based on the analysis of long time periods because the exploration, drilling, and production process takes place over decades; therefore, the recent changes cannot be given too much weight for future planning. For this reason, most long-run models for predicting oil prices are based on considerations of supply, demand, economic trends, new technologies, population growth, etc. These models can (and should) be stochastic, often utilizing techniques we have covered in other chapters (e.g., regression analysis). Rather than choosing between a short-run forecasting model, based on time series analysis, and a long-run model of the fundamental forces at work in the oil industry, we will show an example of how these two approaches may be combined in a single analysis.

We will fit a time series to the 2005–2010 data, predict the next 12 months of oil prices based on this historical pattern, and then consider how to incorporate a shock to the oil market that is not part of the historical record.* Figure 6.24 shows that the GARCH model fits the 2005–2010 data best (according to the Information Criteria), and it appears to give reasonable forecasts for the next year.†

* We do not recommend relying on a time series model based purely on historical data for longer than this period without attempting to model changes in the underlying market conditions.

† There does not appear to be any seasonality in the oil prices. This is not surprising since there are well developed futures markets and oil is a storable commodity (albeit at a cost). We will let the time series methods pick up whatever meandering pattern exists within the data, rather than trying to remove the seasonality or trends.

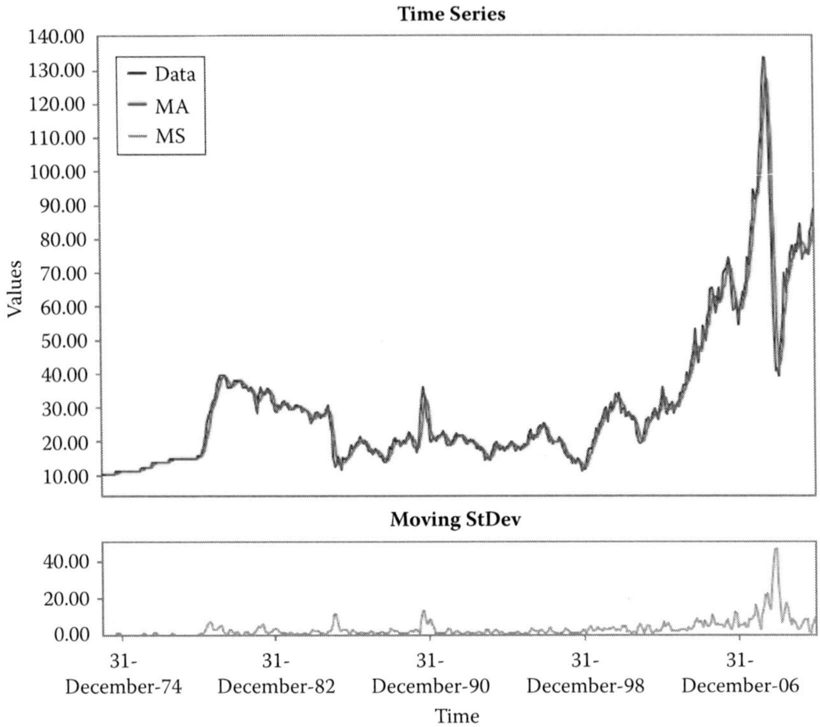

FIGURE 6.23
Oil prices 1974–2010.

We will paste the 12 forecast months after the historical data series (D793:D804), using the December 2010 price (89.04) as the last value. We then create a range of output cells for our 12-month forecast period (E793:E804).* The trend chart, Figure 6.25, suggests that the median oil price is expected to rise very slightly over the next 12 months.

One month into the future, the range of prices is fairly narrow, but by the time a year passes, the forecast range extends from $50 to over $200 per barrel (for a 98% confidence interval). This should be expected, given the extreme volatility exhibited by oil prices, even over fairly short periods of time. Given that oil company decisions span decades of time, it is not surprising that the companies use price forecasts similar to the historical long-run average prices in their planning, even if prices may currently be high.

* This is necessary because we cannot designate an Excel array function as a ModelRisk output range.

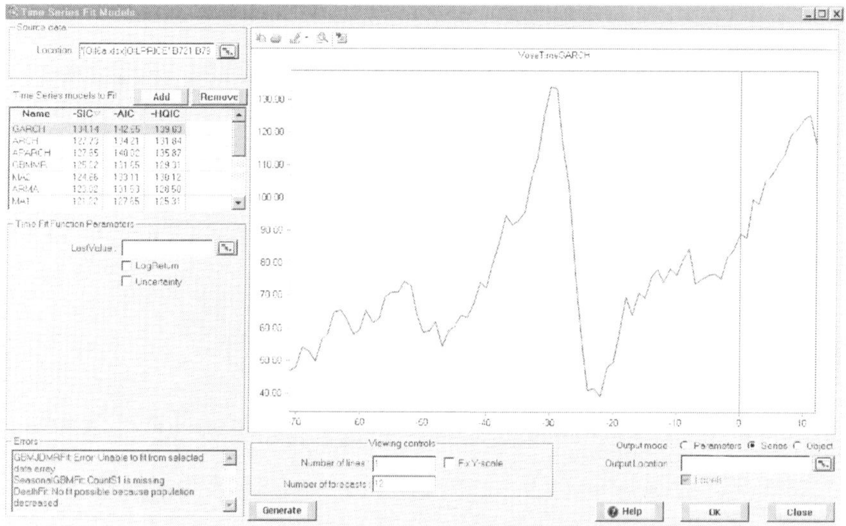

FIGURE 6.24
GARCH model for oil prices.

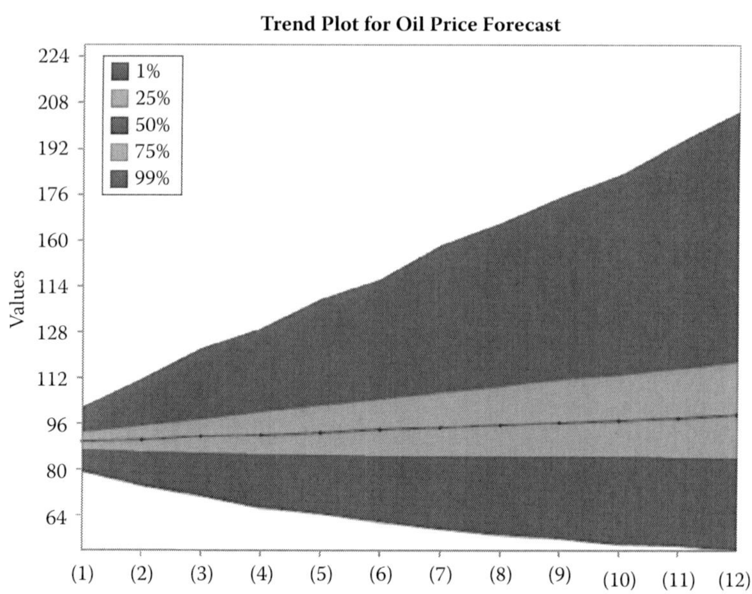

FIGURE 6.25
Oil price trend chart.

Thus far, the forecasting we have discussed within this chapter has been mainly data driven, but we can also use hybrid forecasting models to combine the data-driven time series with expert opinion. The results from the GARCH model are a product of the past historical data used to develop the model. The potential future impacts of the Gulf of Mexico oil disaster in 2010 are not reflected in these data. One potential impact might be a reduction in production from offshore wells due to concerns about further spills.* Indeed, data on oil futures (e.g., 1-, 5-, or 10-year futures contracts) may reveal useful data for calibrating our model. (We do not examine this here.)

We will assume that the supply shock (in absolute value) follows a PERT distribution, with a minimum value of 0%, most likely value of 5%, and maximum value of 10%. The demand elasticity is also uncertain, and we use a PERT distribution with (absolute) values of (0.1, 0.4, 0.5). More extensive research and/or statistical modeling could be used to refine the estimates for the demand elasticity, including application of the regression modeling techniques discussed in Chapter 5. We use these PERT distributions here simply to illustrate how to build a hybrid time series forecasting/judgmental simulation model.

The modeling section of the spreadsheet Oil6b.xlsx is shown in Figure 6.26. Cells B74:B85 contain our GARCH time series forecast. D73 simulates the supply shock, F73 simulates the demand elasticity, and G73 combines these to estimate the price shock that results. In D74:D85 we add these price shocks to the basic GARCH time series forecast from the historical data. We have gradually introduced the price shock over the 12-month period, with the full impact felt in 12 months but proportionally less impact over the preceding 11 months. Before running the simulation, we will designate B86 (=B85) and D86 (=D85) as output cells so that we can compare the results for the 12-month oil price with and without including the supply shocks. Running the simulation provides the comparison shown in Figure 6.27.

Including the supply shock leads to a forecast of generally increased prices, as well as increased uncertainty. The mean price at 12 months in the future increases from $103 to $118 per barrel and the standard deviation increases from $31 to $36.

This model illustrates how to combine historical data-driven classical time series forecasting methods with more subjective models to simulate ways that past patterns may be impacted by new events.

* This concern will more likely impact future drilling, which might not show up in oil prices immediately. However, since oil production can easily be shifted over time, any expected future price increase will also impact short-run prices as well. In any case, we are modeling a short-run impact on oil supply in order to show how to use a forecast model as part of a hybrid model, combined with more judgmental factors, to make predictions.

	A	B	C	D	E	F
72	40483	84.14		Supply shock	Demand elasticity	Price shock (12 month)
73	40513	89.04	time	=VosePERT(0%,5%,10%)	=VosePERT(0.1,0.4,0.5)	=D73/E73
74	Base Forecasts	=VoseTimeGARCHFit(B2:B73,FALSE,FALSE,B73)	1	=B74*(1+(C74/12)*$F$73)		
75		=VoseTimeGARCHFit(B2:B73,FALSE,FALSE,B73)	2	=B75*(1+(C75/12)*$F$73)		
76		=VoseTimeGARCHFit(B2:B73,FALSE,FALSE,B73)	3	=B76*(1+(C76/12)*$F$73)		
77		=VoseTimeGARCHFit(B2:B73,FALSE,FALSE,B73)	4	=B77*(1+(C77/12)*$F$73)		
78		=VoseTimeGARCHFit(B2:B73,FALSE,FALSE,B73)	5	=B78*(1+(C78/12)*$F$73)		
79		=VoseTimeGARCHFit(B2:B73,FALSE,FALSE,B73)	6	=B79*(1+(C79/12)*$F$73)		
80		=VoseTimeGARCHFit(B2:B73,FALSE,FALSE,B73)	7	=B80*(1+(C80/12)*$F$73)		
81		=VoseTimeGARCHFit(B2:B73,FALSE,FALSE,B73)	8	=B81*(1+(C81/12)*$F$73)		
82		=VoseTimeGARCHFit(B2:B73,FALSE,FALSE,B73)	9	=B82*(1+(C82/12)*$F$73)		
83		=VoseTimeGARCHFit(B2:B73,FALSE,FALSE,B73)	10	=B83*(1+(C83/12)*$F$73)		
84		=VoseTimeGARCHFit(B2:B73,FALSE,FALSE,B73)	11	=B84*(1+(C84/12)*$F$73)		
85		=VoseTimeGARCHFit(B2:B73,FALSE,FALSE,B73)	12	=B85*(1+(C85/12)*$F$73)		
86	12 month forecast	=VoseOutput("Forecast without supply shock")+B85		=VoseOutput("Forecast with supply shock")+D85		

FIGURE 6.26
Hybrid oil price model.

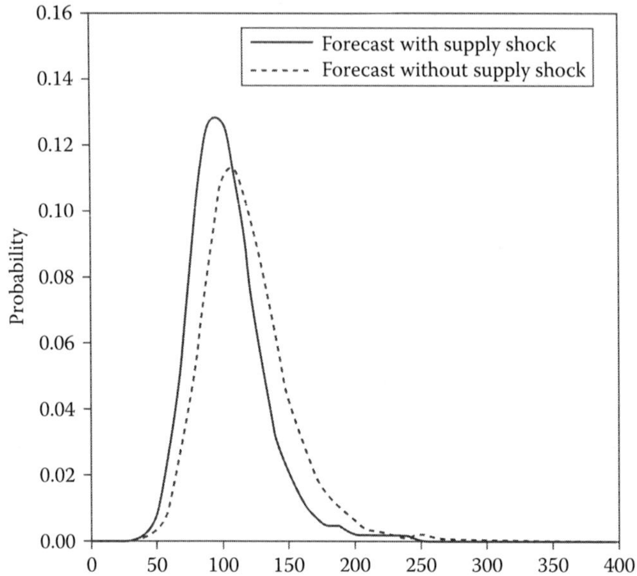

FIGURE 6.27
Hybrid model results.

6.7 Fourth Example: Home Prices and Multivariate Time Series

In Chapter 5 we looked at the importance of modeling relationships between uncertain variables that are represented by probability distributions. In similar fashion, we must be aware that there can also be a relationship between random variables being modeled by a time series. These multiple time series, while fluctuating over time (due to randomness, trends, and seasonality), may do so in a related fashion. For example, we now know that geographic housing markets are somewhat related to one another. The relationship is not causal: It makes no sense to think of housing price changes in Detroit causing house prices to change in Las Vegas (or vice versa). Home prices6.xlsx has data that we have previously seen for monthly housing prices in four U.S. cities (Detroit, Las Vegas, New York, and Dallas). A time structure to these housing prices appears to be similar in each city. (At least two trends are evident: an upward trend in the first half of the data and a downward trend thereafter.) We wish to model these related time series, and for that we use a *multivariate time series model* in order not only to include the time series patterns, but also to capture the relationship between home prices in the four cities.

Using the data viewer to look at these data gives Figure 6.28. The related trends are evident, as is significant correlation between each pair of cities. To

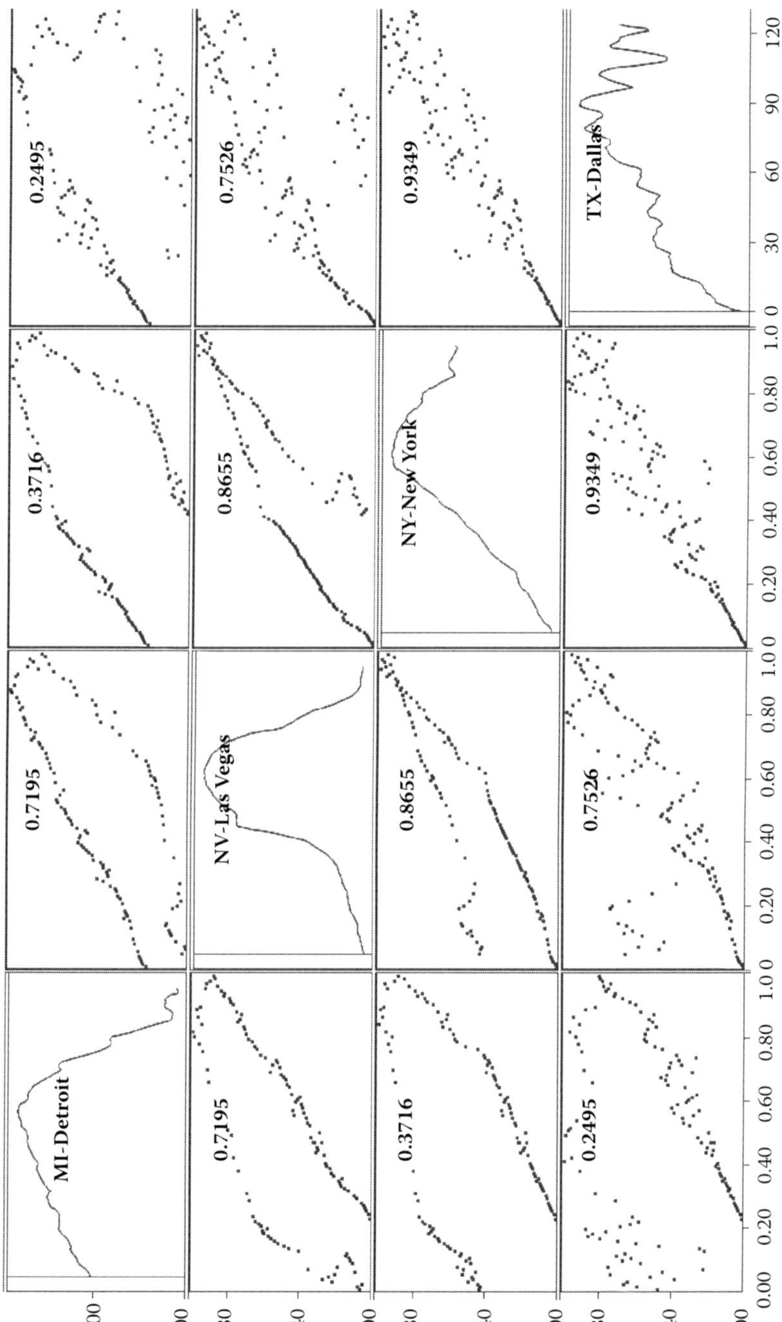

FIGURE 6.28
Multivariate city house price data.

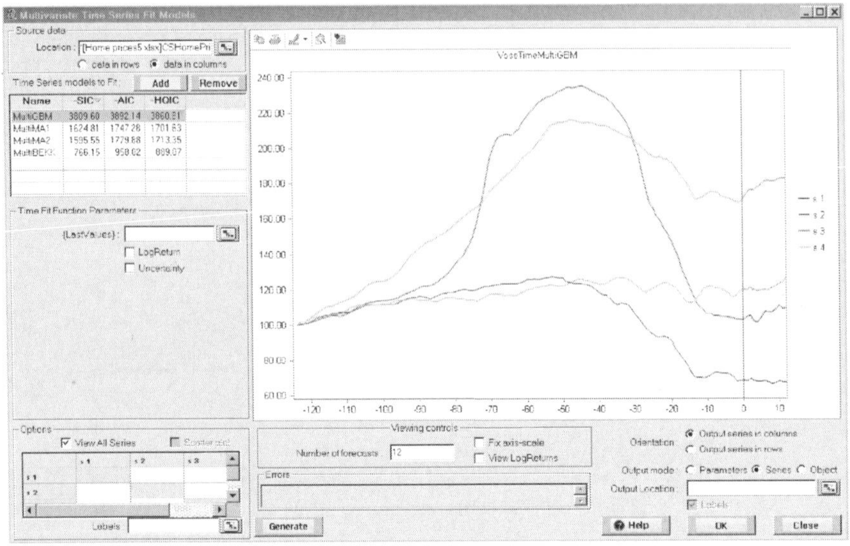

FIGURE 6.29
Multivariate time series models.

analyze these time series, we fit a multivariate time series model from the ModelRisk toolbar and obtain Figure 6.29.

Of the four models available, the Multivariate Geometric Brownian Motion (MultiGBM) model appears to be the best fit. We will use the MultiGBM to forecast 12 months of housing prices and paste these series into the data set. We will use the 1-year forecast from each city to produce simulation outputs. Figure 6.30 shows overlay results for the four city forecasts after running the simulation.

It is not obvious from Figure 6.30, however, that using the multivariate time series model is more than simply an expedient way to fit time series models to each of the four cities. The multivariate time series fit also models the correlation between the time series. In the Simulation Results window, we can choose the scatter plot button and select any two forecasts to see their correlation patterns. Figure 6.31 shows the results for Detroit and New York.

The power of multivariate time series models is that they can easily estimate both the time structure and correlation between these series. This would be very important, for example, for an investor who has real estate holdings in several states and is interested in the risks of house price changes on his or her total holdings. The positive correlation between price changes makes the portfolio more risky, and ignoring the correlation would cause an underestimation of total risk.

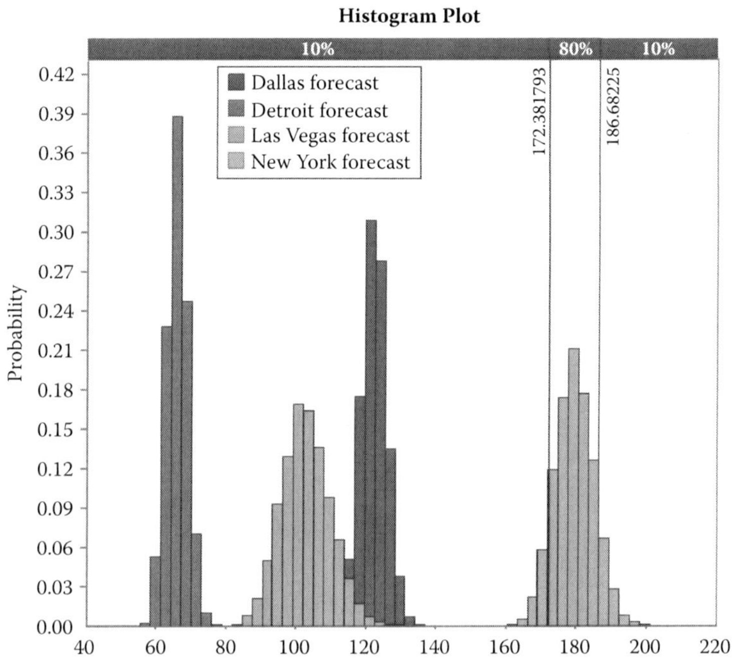

FIGURE 6.30
Overlay of four cities' home price forecasts.

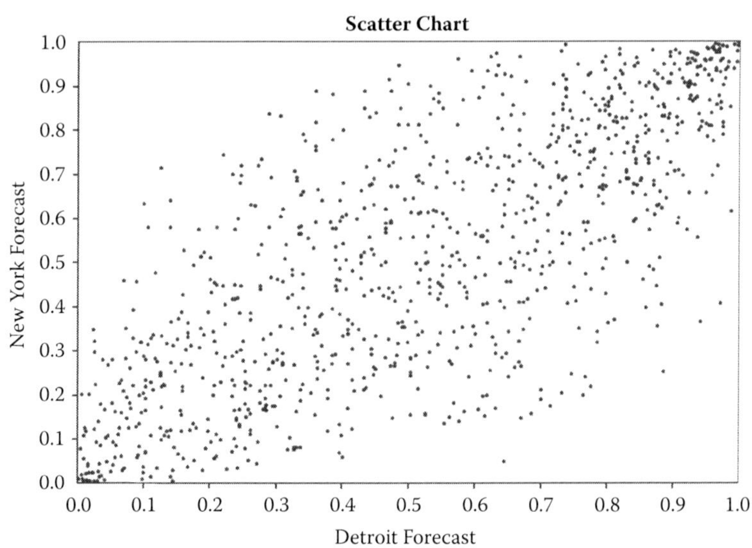

FIGURE 6.31
Detroit versus New York simulation results.

6.8 Markov Chains*

A special, but commonly encountered, type of relationship occurs when considering a system with a random variable that can take on one of several discrete states (e.g., health status, credit ratings, or the state of the economy). The probability of moving from one state to another is dependent only on the current state and is not dependent on any historical transitions. In other words, where the random variable goes next does not in any way depend on where it has been. For example, there may be a given probability of contracting a disease that depends upon whether an individual was previously ill or healthy (but **not** on any of the disease history before the current status). Similarly, there may be a probability of a mortgage holder defaulting on a loan that depends on his or her credit score (but not the entire history of credit scores). This type of system is often called *"memoryless."* When the probability of transitioning from one state to another is constant over time, such problems can be modeled as a *Stochastic Markov Process.*

As a simple example, suppose that you are considering purchasing the account receivables from a wholesale supplier. The firm has 100 accounts and each is invoiced every month. Of the 100 accounts, 60 are currently paid on time in full (p), 20 are 1 month overdue (1), 20 are 2 months overdue (2), and none are in default (bad: b). After an account is 2 months overdue, it is either paid in full or declared bad and sent to collections. You wish to simulate the evolution of these 100 accounts over the next 12 months, given the following transition probabilities:

- There is a 90% probability that a paid account will continue to be paid in full the next month; there is a 10% probability that it will become 1 month overdue.
- A 1-month overdue account has an 85% chance of being paid in full and a 15% chance of becoming 2 months overdue.
- A 2-month overdue account has only a 30% chance of being paid in full, but a 70% chance of being sent to collections.
- A bad account has a 100% chance of remaining a bad account.†

These probabilities can be represented in a transition matrix, shown in Figure 6.32.

The entries in the matrix show the probability of moving from any one state to another state in the next time period. The structure of this matrix can be made more complicated—for example, by making the transition

* This section is more advanced (though not difficult) and can be skipped without loss of continuity.

† Such a state is called an *absorption state*; once entered, it will not be left.

		To:			
		p	1	2	*b*
From	*p*	0.9	0.1	0	0
	1	0.85	0	0.15	0
	2	0.3	0	0	0.7
	b	0	0	0	0

FIGURE 6.32
Transition matrix.

probabilities themselves uncertain (which, in reality, they typically are). An important requirement is that all the transition probabilities in every row sum to one. Nested logical statements (e.g., IF(AND), IF(OR), etc.) can be used to model the transition from one period to the next. While it is fairly straightforward to model the transition of the 100 accounts to the next month, it quickly becomes very cumbersome within Excel to model the evolution of a large number of states or a large number of time periods.

ModelRisk contains a feature that easily models a Markov Chain. Markov6.xlsx shows such a model for the 100 accounts receivable. Figure 6.33 shows part of the model. (The model spans 12 months, but only the formulas for months 0 and 1 are shown in full.)

Column H (rows 5 through 8) show the initial set of accounts in each state. Each column (e.g., cells C14:C17, D14:D17, etc.) uses an array function, VoseMarkovSample, which references the initial set of accounts, the transition matrix, and the number of time periods (transitions). Thus, column C shows the initial data (zero transitions), D shows one transition, all the way to column N (12 months or transitions), as shown in Figure 6.34 (showing the cell results, rather than the formulas).

Rows 20 through 25 simply reproduce the values from rows 12 through 17; this is to permit ModelRisk to display the simulation results.* We have highlighted the entire rows for the paid accounts and bad accounts (rows 22 and 25) and marked them as ModelRisk outputs, entering "paid accounts" and "bad accounts" as *range names,* as shown in Figure 6.35.

This will permit the simulation results to be displayed as a *trend chart.* Selecting F9 will show a single iteration of the initial 100 accounts through 12 months of transitions. Running 10,000 simulations produces the trend charts for paid accounts and bad accounts shown in Figures 6.36 and 6.37.

Each trend chart shows the median simulation, a 50% prediction interval, and a 98% interval for each month's number of accounts. (Which exact

* It is not possible to modify parts of an array formula in Excel; inserting the ModelRisk output marking would be such a modification, and attempting to do so would cause Excel to generate an error message. Remember that to enter the array function, first highlight the range of cells (e.g., C14:C17), enter the function (e.g., VoseMarkovSample), and press Ctrl + Shift + Enter). Then, column C can be copied through to column O.

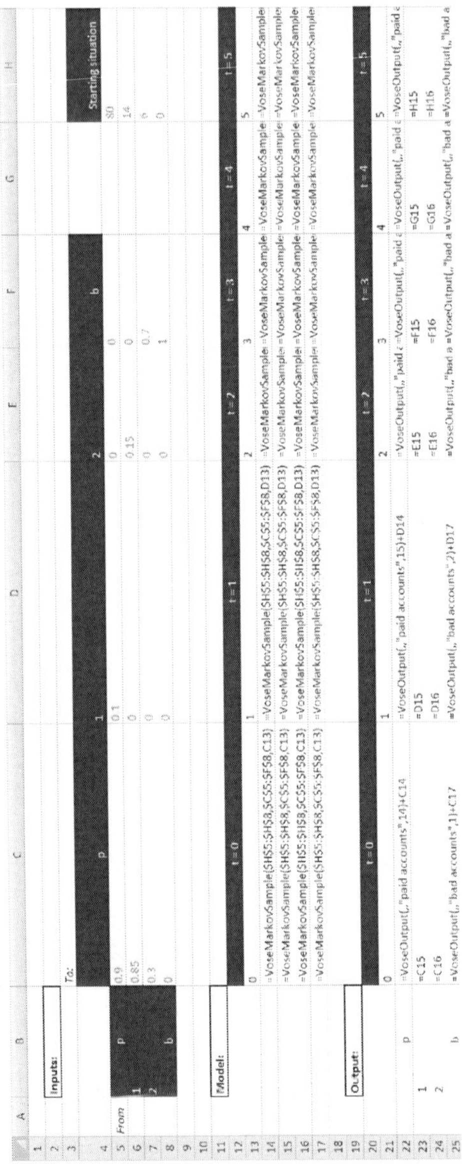

	A	B	C	D	E	F	G	H
1								
2		Inputs:						
3			To:					
4			p	1	2	b		Starting situation
5	From	p	0.9	0.1	0	0		80
6		1	0.85	0	0.15	0		14
7		2	0.3	0	0	0.7		6
8		b	0	0	0	1		0
9								
10								
11		Model:						
12			t = 0	t = 1	t = 2	t = 3	t = 4	t = 5
13			0	1	2	3	4	5
14			=VoseMarkovSample($H$5:$H$8,$C$5:$F$8,C13)	=VoseMarkovSample($H$5:$H$8,$C$5:$F$8,D13)	=VoseMarkovSample	=VoseMarkovSample	=VoseMarkovSample	=VoseMarkovSample
15			=VoseMarkovSample($H$5:$H$8,$C$5:$F$8,C13)	=VoseMarkovSample($H$5:$H$8,$C$5:$F$8,D13)	=VoseMarkovSample	=VoseMarkovSample	=VoseMarkovSample	=VoseMarkovSample
16			=VoseMarkovSample($H$5:$H$8,$C$5:$F$8,C13)	=VoseMarkovSample($H$5:$H$8,$C$5:$F$8,D13)	=VoseMarkovSample	=VoseMarkovSample	=VoseMarkovSample	=VoseMarkovSample
17			=VoseMarkovSample($H$5:$H$8,$C$5:$F$8,C13)	=VoseMarkovSample($H$5:$H$8,$C$5:$F$8,D13)	=VoseMarkovSample	=VoseMarkovSample	=VoseMarkovSample	=VoseMarkovSample
18								
19		Output:						
20			t = 0	t = 1	t = 2	t = 3	t = 4	t = 5
21			0	1	2	3	4	5
22		p	=VoseOutput(,,"paid accounts",14)+C14	=VoseOutput(,,"paid accounts",15)+D14	=VoseOutput(,,"paid a	=VoseOutput(,,"paid a	=VoseOutput(,,"paid a	=VoseOutput(,,"paid a
23		1	=C15	=D15	=E15	=F15	=G15	=H15
24		2	=C16	=D16	=E16	=F16	=G16	=H16
25		b	=VoseOutput(,,"bad accounts",1)+C17	=VoseOutput(,,"bad accounts",2)+D17	=VoseOutput(,,"bad a	=VoseOutput(,,"bad a	=VoseOutput(,,"bad a	=VoseOutput(,,"bad a

FIGURE 6.33
Markov model construction.

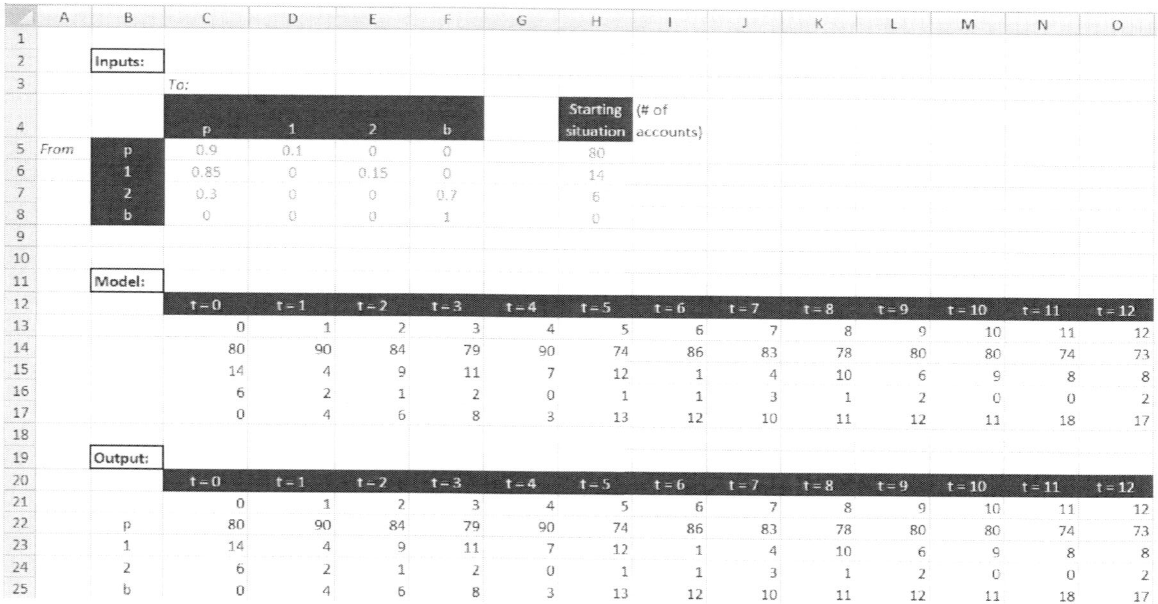

	A	B	C	D	E	F	G	H	I	J	K	L	M	N	O
1															
2		Inputs:													
3			To:												
4			p	1	2	b		Starting situation	(# of accounts)						
5	From	p	0.9	0.1	0	0		80							
6		1	0.85	0	0.15	0		14							
7		2	0.3	0	0	0.7		6							
8		b	0	0	0	1		0							
9															
10															
11		Model:													
12			t = 0	t = 1	t = 2	t = 3	t = 4	t = 5	t = 6	t = 7	t = 8	t = 9	t = 10	t = 11	t = 12
13			0	1	2	3	4	5	6	7	8	9	10	11	12
14			80	90	84	79	90	74	86	83	78	80	80	74	73
15			14	4	9	11	7	12	1	4	10	6	9	8	8
16			6	2	1	2	0	1	1	3	1	2	0	0	2
17			0	4	6	8	3	13	12	10	11	12	11	18	17
18															
19		Output:													
20			t = 0	t = 1	t = 2	t = 3	t = 4	t = 5	t = 6	t = 7	t = 8	t = 9	t = 10	t = 11	t = 12
21			0	1	2	3	4	5	6	7	8	9	10	11	12
22		p	80	90	84	79	90	74	86	83	78	80	80	74	73
23		1	14	4	9	11	7	12	1	4	10	6	9	8	8
24		2	6	2	1	2	0	1	1	3	1	2	0	0	2
25		b	0	4	6	8	3	13	12	10	11	12	11	18	17

FIGURE 6.34
Markov model.

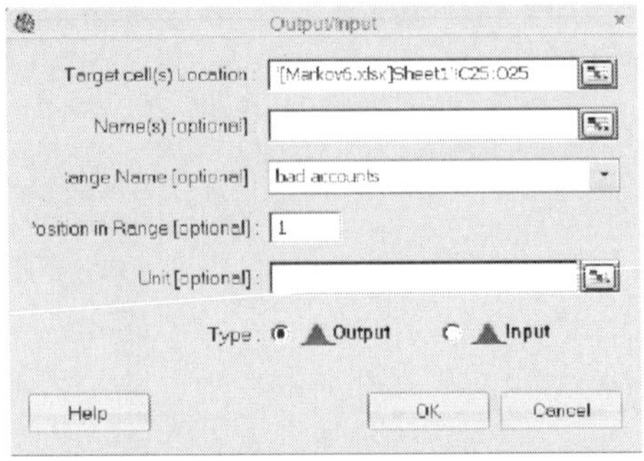

FIGURE 6.35
ModelRisk output range for trend chart.

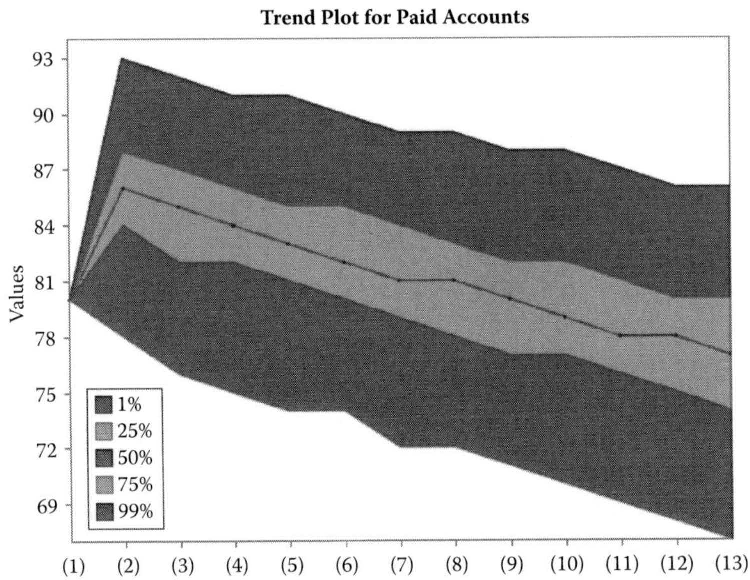

FIGURE 6.36
Trend chart for paid accounts.

confidence interval to display can be adjusted manually.) The number of bad accounts (which starts at zero) evidently should be expected to climb over time. After 1 year, we would expect 13 bad accounts, with a 1% chance of as many as 22 bad accounts. The number of paid accounts should rise in the first month (as more 1-month-overdue accounts are paid in full than paid accounts become overdue), but is likely to fall afterward.

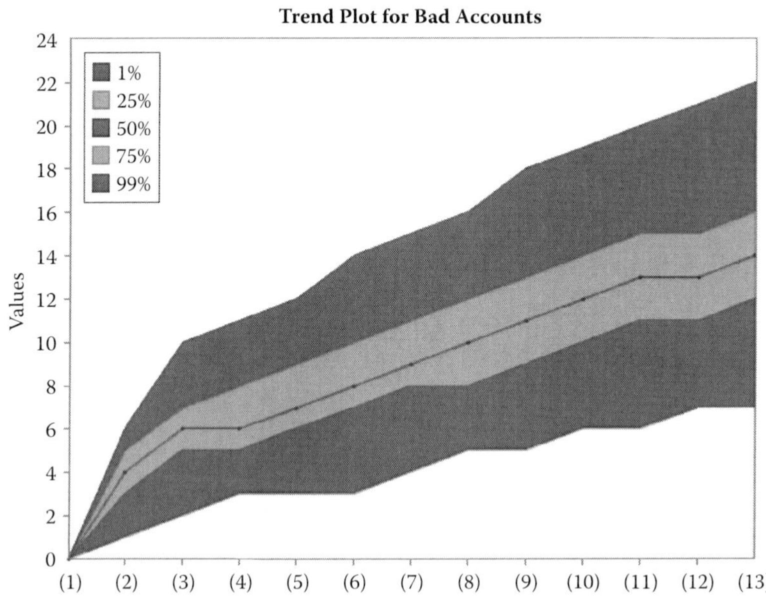

FIGURE 6.37
Trend chart for bad accounts.

6.9 Summary

The majority of the time series forecasting models that we have discussed are data driven. They model whatever structure or patterns are present in historical data, incorporating the level, trend, seasonality, volatility, and randomness of the time series in order to predict future values. They should only be used if the underlying process is stable and not expected to change significantly in the future. If the underlying process has been stable, but may change in the future, then the data-driven forecasting methods can be combined with more judgmental simulation in a hybrid approach, or, alternatively, a forecast can be based on expert opinion alone.

In this chapter, we have mostly focused on time series methods that are based on historical data, but they can also be designed in the absence of data. An example would be a model for retail sales of a new product, in which case there is likely no good historical data available and we would likely use expert opinion.

CHAPTER 6
Exercises

6.1 FLU SEASON

Exercise6-1.xlsx contains data on the number of flu cases in the United States from September 28, 2003, through the end of December 2010.* Assume that the flu season starts in the last week of September each year.

a. How would you characterize the 2010–2011 flu season (as of the end of December)—is it above or below average? Does your answer vary for different weeks during the season?
b. Estimate the probability that the total 2010–2011 season will experience more flu cases than did the 2009–2010 season.
c. Estimate the probability that a week during the 2011 calendar year of the 2010–2011 season will experience the largest number of flu cases of all the weeks since the start of the 2003–2004 season.
d. Discuss some of the main assumptions you have made in your analysis. Are they valid? How can the forecast possibly be improved?

6.2 AIRLINE LOAD FACTORS

Airline load factors are a primary driver of airline profitability. (Of course, prices and costs also matter: A full plane will lose money if price is less than average cost and a half-empty flight can be profitable if price is greater than average cost.) Some low-cost airlines may be profitable at load factors as low as 64%, while others may require load factors of 100% or more.† Exercise6-2.xlsx contains data for the U.S. airline industry for passenger load factors for domestic and international operations over the period of 2000 through September 2010.

a. Inspect the data and comment on whether there appear to be any trends, seasonality, or randomness in the two series.
b. Fit a time series model to each series separately and predict the next 10 months of load factors (October 2010–July 2011).
c. Fit a multivariate time series to the two sets of data and compare and discuss your forecast 10 months out with your results for part b.

6.3 STRANGLE OPTIONS

A *long strangle* is a financial strategy aimed to capitalize on volatility in stock prices. It consists of buying a call option and selling a put option on the same security that expire on the same date. The call option gives

---

* Data accessed from www.google.org/flutrends/us/#US on January 4, 2011.

† Vasigh, B., K. Fleming, and L. MacKay. 2010. *Foundations of Airline Finance*. Surrey, England: Ashgate Publishing.

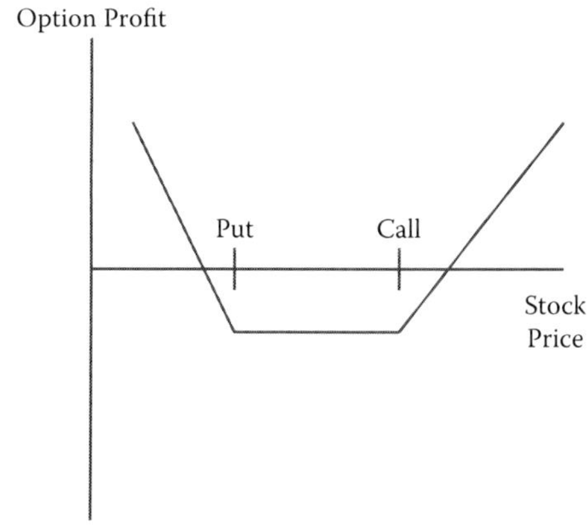

FIGURE 6.38
A long strangle option.

the buyer the right to purchase an underlying asset for a particular price (called the strike price) at a particular point in time. A put option gives the seller the right to sell the asset at a specified price at a particular point in time. Thus, an investor who buys a call option at a higher strike price than he or she sells a put option will profit if the stock price is sufficiently volatile. Figure 6.38 illustrates the long strangle.

When the stock price rises above the call option's strike price ("call" in Figure 6.29), then the option will be exercised and the gain will equal the difference between the actual stock price and the call strike price. When the stock price falls below the put option's stock price ("put" in Figure 6.29), then the option will be exercised and the gain will be the difference between the put strike price and the actual stock price. If the price stays between the two strike prices, then the options will not be exercised and no gain will be realized; in fact, a loss will occur since purchasing these options costs something. Thus, the long strangle is a gamble that a stock price will deviate substantially from its current value.

Exercise6-3.xlsx has daily stock price data for 2008–2010 for AIG International (ticker symbol: AIG). Recent experience with this stock might suggest that a long strangle would be an appropriate strategy.

a. Build a time series model to predict the daily changes in AIG stock for the first 100 days of trading in 2011. Use the percent change column to produce the time series model. Once you have the predicted daily stock price changes, calculate the predicted stock prices by using the fact that each day's stock price = the previous day's price × $e^{\%\text{ daily change}}$.*
b. Consider a strangle option with a call strike price 10% higher than the end-of-year 2010 price and a put strike price 10% lower

* This comes from the fact that the percent change is calculated as ln (pricet/pricet–1) so that pricet = $e^{\%\text{change}}$ pricet–1. Use the EXP function in Excel.

than the end-of-year 2010 price. What is the expected value of this option? In other words, how much should you be willing to pay to engage in this strangle option? Provide a 90% confidence interval for this value, assuming the option is a 25-, 50-, 75-, or 100-day option.

c. Look up the actual AIG stock price 100 days beyond the end of the data provided. If you had paid the expected value of this strangle option, how would you have fared?

6.4 E-COMMERCE AND RETAIL SALES

Exercise6-4.xlsx has quarterly data (1999–2010) for retail sales in total and e-commerce retail sales.* Clearly, e-commerce comprises an increasing portion of retail sales activity.

a. Examine the total retail sales and e-commerce sales data and comment on the relevance of trends, seasonality, and randomness in each series. Does there appear to be any correlation between the two data series?
b. Build a time series model to simulate one additional year of sales (both total and e-commerce) for 2010.† Do univariate and multivariate fits for both data series.
c. What is the probability that e-commerce sales in 4Q 2010 will exceed those from 4Q 2009? What is the probability that e-commerce sales will comprise a greater portion of total retail sales activity in 4Q 2010 than in 3Q 2010?

6.5 PATENTS

Exercise6-5.xlsx contains annual data on the number of patent applications for utility (invention) patents and design patents from 1880 through 2009.‡ You wish to forecast the number of patents of each type that will be applied for and received over the next 10 years.

a. Fit time series models to the patent application data—either univariate or multivariate time series models, as you feel are appropriate. Explain your choice of models.
b. Fit time series models for the fraction of patent applications granted for each type of patent. Again, use either univariate or multivariate methods and explain your choice.
c. Using your results from part (b), simulate the fraction of each type of patent application granted over the next 10 years. Provide confidence intervals for your estimates. (Trend charts would be useful.)

---

* The data come from the U.S. Census Bureau and are measured in millions of dollars.

† While those data are now available, they had not yet been released at the time of this writing. For this question, pretend that it is October 1, 2010.

‡ The data come from the U.S. Patent and Trademark Office.

6.6 GLOBAL WARMING

Global climate issues are important, complex, and controversial, and data sources are extensive. Exercise6-6.xlsx is not meant to be an exhaustive study of the issue, but merely to provide an initial view. In fact, time series forecasting methods should not be applied over time periods this long. Data are provided on the temperature (difference from the average of the last 1,000 years) and CO_2 concentrations over the past 800,000 years. The trends and relationships are debated, and even the way these variables are measured is fraught with controversy.

a. Investigate a time series model for each variable separately, based on all the data except for the latest observation. Predict the current value (the latest 1,000 years' observation) for each, assuming that the historical pattern remains unchanged. *Hint:* it will help to normalize the two series so that they are on a comparable scale and can be investigated visually. You will need to rescale the temperature data anyway since the negative values would be interpreted by ModelRisk as log Return data (percentage changes). From a graph of the data series, comment on the latest observed values.
b. Perform a multivariate time series analysis of the same variables.
c. Estimate the probability of observing the latest temperature anomaly and CO_2 concentration for both the univariate and multivariate models. Discuss the assumptions you made in your analysis and if they seem reasonable.

6.7 NEW CAR SALES

Exercise6-7.xlsx has monthly data on total new car registrations in the EU15 countries from 1990 through September 2010.* You wish to forecast the number of sales for the last 3 months of 2010 and the following 2 years.

a. Visually investigate the time series and comment on trends, seasonality, volatility, and randomness.
b. Fit a time series model to the data and explain your choice of model.
c. Simulate the 27 forecast months of sales. Based on your forecast, would you recommend expanding or reducing the current dealership network in the EU15? Justify your decision with a probabilistic assessment.

6.8 THE OLYMPIC SWIMSUIT CONTROVERSY

On February 13, 2008, the LZR Racer swimsuit, which was claimed to reduce race times by 1%–2%, was introduced. By August 14, 2008, 62 world swimming records had been broken. Exercise6-8.xlsx provides winning times for the Olympic Gold medals in two sports: the men's 100-m freestyle swim and the men's 100-m track-and-field race.

* Data are from ACEA, the European Automobile Manufacturers' Association.

a. Fit appropriate time series models to the two series, using the data from 1896 to 2004. Forecast the winning 2008 times and provide a probabilistic assessment.
b. Based on these data alone, do you think the recent banning of these swimsuits in Olympic competition is justified? Discuss the main assumptions you made and how you might possibly refine your analysis.

7

Optimization and Decision Making

LEARNING OBJECTIVES

- Understand when and where optimization is appropriate and useful.
- Appreciate how to use tables to get approximately optimal solutions.
- Learn how to use OptQuest to find optimal solutions.
- Learn how to combine optimization and simulation.
- Understand the difference between Excel's Solver and OptQuest.
- Understand potential trade-offs between risk and return.
- Understand the differences between constraints and requirements.
- Learn how to use simulation to improve decision making.

7.1 Introductory Case: Airline Seat Pricing

The airline industry has led the revolution in revenue management pricing.* On the whole, airfares are a complex set of fare classes, where each fare class is subject to a myriad of restrictions. Deriving the best set of fare classes to offer for a single route alone is highly complex, but the complexity increases dramatically as the number of route segments grows. The best price for a seat on any particular flight leg typically also depends on the demand for that seat by all potential passengers that might use that leg as part of their journey.

One of the first and simplest solutions to the single-leg pricing problem was proposed by Littlewood.† A search for academic articles using the terms "Littlewood," "pricing," and "revenue" shows 152 published articles, with 130 of them published since the year 2000.‡ Littlewood considered a single flight leg with two fare classes. The demand for the lower fare class is assumed to arrive before the demand for the higher fare class. Littlewood's suggested solution to this pricing problem is basically to accept demand for

* See Boyd. E. A. 2007. *The Future of Pricing: How Airline Ticket Pricing Has Inspired a Revolution.* Hampshire, England: Palgrave Macmillan.

† Littlewood, K. 1972. Forecasting and control of passenger booking. AGIFORS Proceedings, 12th Annual Symposium Proceedings.

‡ Search conducted on Business Source Premier, January 6, 2011.

the lower price class seats until the known definite revenue from selling one more low price seat is exceeded by the expected revenue (which is uncertain) of selling the same seat at the higher price. We will examine whether this principle does indeed provide optimal pricing when the demand for seats is uncertain.

7.2 A Simulation Model of the Airline Pricing Problem

Consider a Wednesday morning flight from Philadelphia to Albany. The aircraft has 100 seats and two fare classes: Y (first class), which sells for $300, and Q (economy class), which sells for $100.* The demand for Y class seats is normally distributed, with a mean of 30 seats and a standard deviation of 10.† Q class demand is also normally distributed, but with a mean of 90 and a standard deviation of 40. In this example, we make the simplifying assumption that the Q class passengers make reservations before the Y class passengers. Thus, the decision problem becomes how many Y class passengers to accept before closing the plane to Q class passengers. If we leave seats empty, there is the potential that there will not be any Y class passengers to demand them, but if we fill the seats with Q class passengers, then we may not have seats for the high-paying Y class passengers. Also, assume that any unmet demand due to lack of seats will be lost to other airlines.

Following Boyd's paper,‡ we examine three strategies:

- Strategy 1: no control—passengers are permitted to purchase tickets as long as seats are available.
- Strategy 2: mean control—since we expect an average of 30 Y class passengers, restrict Q class purchases to no more than 70. Clearly this is the optimal strategy if the demand levels are certain (which they are not) and equal to their means. (Then 30 Y class seats and 70 Q class seats will always be sold.)
- Strategy 3: Littlewood's Rule—restrict Q class purchases so that the *expected* incremental revenue for each fare class is equalized. Our initial spreadsheet will examine this as restricting Q class purchases to no more than 63 seats (which appears in Boyd), but we will see that it is not a correct statement of Littlewood's Rule.

* This is the same hypothetical problem examined in Boyd (2007).

† We will truncate these distributions to ensure that demand is never negative; since the Normal distribution is continuous, there is a (small) probability of generating a negative demand from these Normal distributions. We could use alternative distributions, but we will stay close to the problem discussed in Boyd (2007).

‡ Boyd. E. A. 2007. *The Future of Pricing: How Airline Ticket Pricing Has Inspired a Revolution.* Hampshire, England: Palgrave Macmillan.

	A	B	C	D	E	F	G	H	I	J	K	L	M	N	O
1	Seats	100		price	Demand		reserved	30			37				
2			Y class	$300	VoseNormal(30,10,,VoseXBounds(0,))										
3			Q class	$100	VoseNormal(90,40,,VoseXBounds(0,))										
4					Strategy 1			Strategy 2			Strategy 3			Empty Seats	
5	Week	Q demand	Y demand	Q sales	Y sales	Revenue1	Q sales	Y sales	Revenue2	Q sales	Y sales	Revenue3	Strategy1	Strategy2	Strategy3
6	1	147	30	100	0	$10,000	70	30	$16,000	63	30	$15,300	0	0	7
7	2	75	36	75	25	$15,000	70	30	$16,000	63	36	$17,100	0	0	1
8	3	120	49	100	0	$10,000	70	30	$16,000	63	37	$17,400	0	0	0
9	4	125	40	100	0	$10,000	70	30	$16,000	63	37	$17,400	0	0	0
10	5	95	44	95	5	$11,000	70	30	$16,000	63	37	$17,400	0	0	0
11	6	112	40	100	0	$10,000	70	30	$16,000	63	37	$17,400	0	0	0
12	7	99	10	99	1	$10,200	70	10	$10,000	63	10	$9,300	0	20	27
13	8	130	23	100	0	$10,000	70	23	$13,900	63	23	$13,200	0	7	14
14	9	155	37	100	0	$10,000	70	30	$16,000	63	37	$17,400	0	0	0
15	10	123	35	100	0	$10,000	70	30	$16,000	63	35	$16,800	0	0	2
16	11	145	35	100	0	$10,000	70	30	$16,000	63	35	$16,800	0	0	2
17	12	28	43	28	43	$15,700	28	43	$15,700	28	43	$15,700	29	29	29
18	13	29	53	29	53	$18,800	29	53	$18,800	29	53	$18,800	18	18	18
19	14	116	30	100	0	$10,000	70	30	$16,000	63	30	$15,300	0	0	7
20	15	141	18	100	0	$10,000	70	18	$12,400	63	18	$11,700	0	12	19
21	16	61	52	61	39	$17,800	61	39	$17,800	61	39	$17,800	0	0	0
22	17	97	27	97	3	$10,600	70	27	$15,100	63	27	$14,400	0	3	10
23	18	78	44	78	22	$14,400	70	30	$16,000	63	37	$17,400	0	0	0
24	19	146	31	100	0	$10,000	70	30	$16,000	63	31	$15,600	0	0	6
25	20	132	26	100	0	$10,000	70	26	$14,800	63	26	$14,100	0	4	11
26	21	150	26	100	0	$10,000	70	26	$14,800	63	26	$14,100	0	4	11
27	22	66	34	66	34	$16,800	66	34	$16,800	63	34	$16,500	0	0	3
28	23	65	16	65	16	$11,300	65	16	$11,300	63	16	$11,100	19	19	21
29	24	84	51	84	16	$13,200	70	30	$16,000	63	37	$17,400	0	0	0
30	25	95	43	95	5	$11,000	70	30	$16,000	63	37	$17,400	0	0	0
31	26	93	25	93	7	$11,400	70	25	$14,500	63	25	$13,800	0	5	12
32	27	91	19	91	9	$11,800	70	19	$12,700	63	19	$12,000	0	11	18
33	28	152	30	100	0	$10,000	70	30	$16,000	63	30	$15,300	0	0	7
34	29	70	36	70	30	$16,000	70	30	$16,000	63	36	$17,100	0	0	1
35	30	81	43	81	19	$13,800	70	30	$16,000	63	37	$17,400	0	0	0
36	Total					$358,800			$460,600			$468,400	66	132	226

FIGURE 7.1
Airline pricing model.

We will simulate 30 weeks of revenue from these three strategies. The model first simulates the demand for Q and Y seats and then calculates the revenues for the three strategies. Figure 7.1 shows our initial model.

We use Objects for the demand distributions (E2:E3), so the parameter section of the model is quite simple. Columns B and C simulate the weekly demand for Q seats and Y seats. For each strategy, the Q sales are determined by comparing the demand with the seats available for Q class and using whatever is smaller.* Y sales are then constrained by the available seats minus those that have been sold to Q class passengers (which is why Strategy 1, no control, shows many weeks where no Y sales occur). To ensure that demand does not take on negative values, we truncate the Normal distributions as illustrated in Figure 7.2.

The presence of VoseXBounds(0,) in the demand distribution Objects reflects the fact that we have truncated the distribution to exclude negative values. Given the two prices and two demands, Airline pricing7a.xlsx examines the three pricing strategies described previously: no reserved seats (saved for Q class customers), 30 reserved seats, and 37 reserved seats.

* Figure 7.1 shows the demand levels as integers, but in reality the Normal distribution is producing continuous values. We have chosen to round off the demand levels to integer values (both in the display and when calculating sales) to clarify the exposition. Fractional values make sense if we interpret these demand levels as average weekly demands. It would be more appropriate to model demand using a discrete distribution such as the Poisson or Polya distribution, but we are using the same assumptions as Boyd (2007).

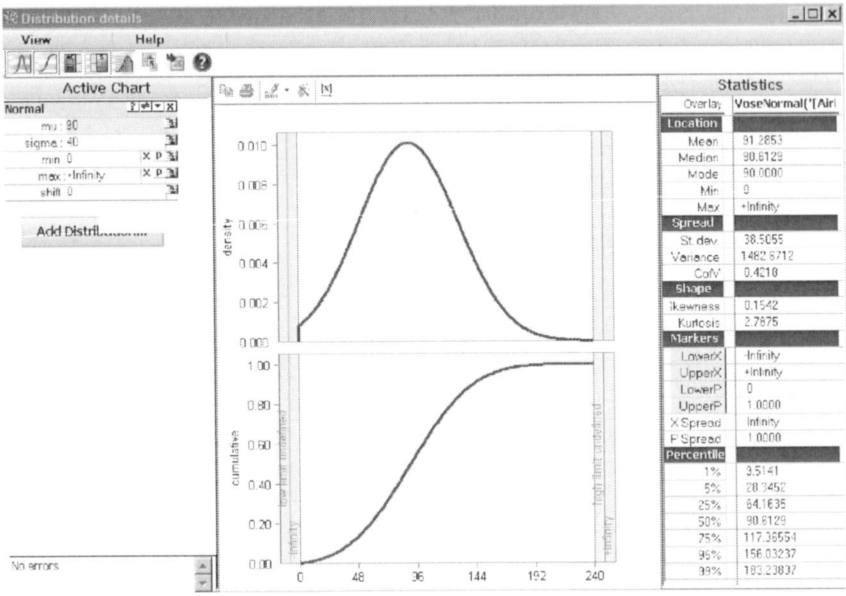

FIGURE 7.2
Truncating a distribution.

Figure 7.3 shows the formulas of the model for the first 4 simulated weeks of Strategy 1 and Strategy 2.

Columns B and C simply simulate the uncertain demand for each fare class for each of the 30 weeks in the model. Sales of Q class seats are then compared with the limit set by the strategy. (No control provides 100 possible seats, cell B1, and mean control reserves 30 seats for Y class, so only 70 are available, B1 minus H1, for Q class demand.) If the demand for Q class is less than the available seats, then the demanded number will be sold, but if it is greater, then only the seats available will be sold. For Y class, if the demand is less than the seats remaining (after Q class passengers have bought their seats), then the demanded number will be sold; otherwise, it will be the lesser of the demand or the seats remaining. For example, suppose that there is demand for 40 Y class seats and that 70 Q class seats have already been sold. Only 30 seats remain for the Y class passengers. However, if the same 40 Y class seats are demanded, but only 55 Q class passengers bought tickets, then only 40 Y class tickets will be sold, even though there are 45 seats available. See cell H6 to view how this logic appears in the Excel formula.

Once we simulate the level of Q and Y class sales, it is straightforward to calculate the total revenue for the flight. We calculate the total for the 30 weeks under each strategy and designate these as simulation outputs. We also keep track of how many empty seats are on each flight and mark those totals as outputs. Running the model and overlaying the total revenue box plots gives Figure 7.4.

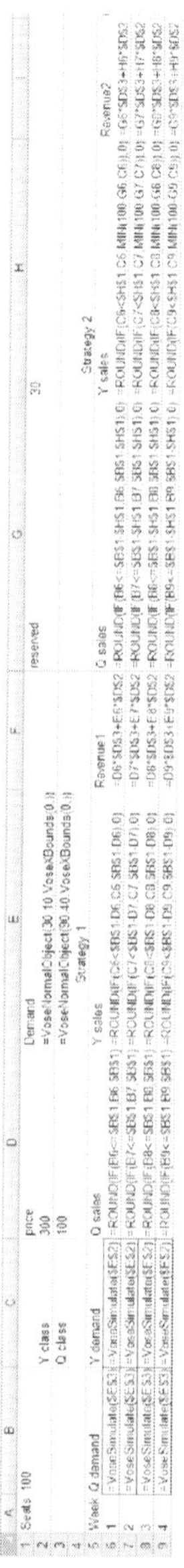

FIGURE 7.3
Four weeks of formulas.

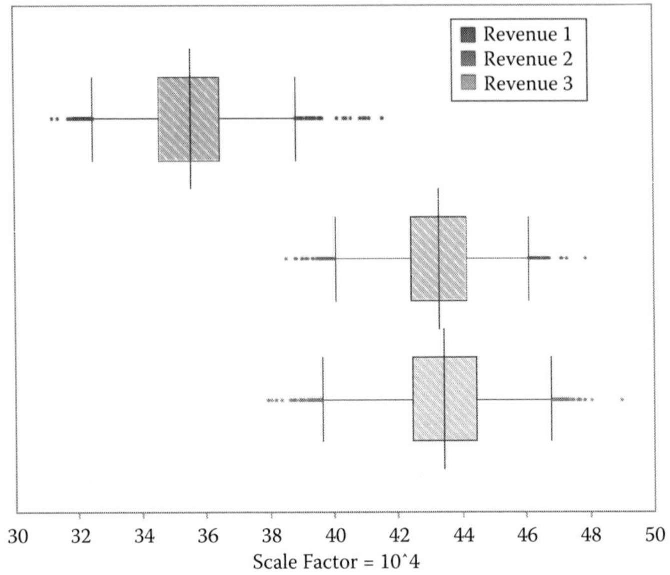

FIGURE 7.4
Revenue results for the three strategies.

The no-control strategy (Strategy 1) is clearly not the best approach, but the mean-control strategy and Littlewood's Rule strategy appear to produce similar results. The median revenue for Littlewood's Rule (Strategy 3) is slightly greater than for the mean-control strategy (Strategy 2), although it appears to be somewhat more variable.* The results for the number of empty seats under each strategy are shown in Figure 7.5.

Strategy 3 has more empty seats than Strategy 2, but earns slightly higher revenues. This is due to the increased number of Y class passengers that it allows. But, the question is whether Littlewood's Rule produces the best result under the assumptions of this model. That is, is the cutoff of 63 Q class seats (37 Y class seats) the best cutoff under these conditions?

7.3 A Simulation Table to Explore Pricing Strategies

A simulation table is a possible way to address this question. Figure 7.6 shows the setup of the simulation table (now included in the model file Airline pricing7b.xlsx). The cells with VoseSimMean and VoseSimStdev

* Strategy 2 generates 21.7% more mean revenue that Strategy 1, while Strategy 3 generates 22.1% more revenue. This compares with Boyd's results of 21.4 and 24.6%, respectively. As we will see, Boyd's application of Littlewood's Rule is not quite accurate.

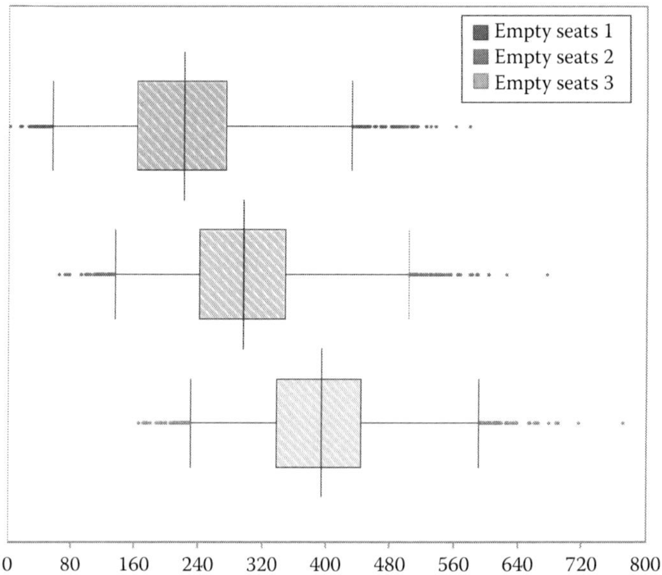

FIGURE 7.5
Empty seat results.

	P	Q	R	S
1	sim #	reserved	expected total revenue	std dev
2	1	30	=VoseSimMean($L$36,P2)	=VoseSimStdev($L$36,P2)
3	2	31	=VoseSimMean($L$36,P3)	=VoseSimStdev($L$36,P3)
4	3	32	=VoseSimMean($L$36,P4)	=VoseSimStdev($L$36,P4)
5	4	33	=VoseSimMean($L$36,P5)	=VoseSimStdev($L$36,P5)
6	5	34	=VoseSimMean($L$36,P6)	=VoseSimStdev($L$36,P6)
7	6	35	=VoseSimMean($L$36,P7)	=VoseSimStdev($L$36,P7)
8	7	36	=VoseSimMean($L$36,P8)	=VoseSimStdev($L$36,P8)
9	8	37	=VoseSimMean($L$36,P9)	=VoseSimStdev($L$36,P9)
10	9	38	=VoseSimMean($L$36,P10)	=VoseSimStdev($L$36,P10)
11	10	39	=VoseSimMean($L$36,P11)	=VoseSimStdev($L$36,P11)
12	11	40	=VoseSimMean($L$36,P12)	=VoseSimStdev($L$36,P12)

FIGURE 7.6
Simulation table setup.

in the formula will show "No simulation results" until the simulation has been run. The other step for running the SimTable is to change Strategy 3 so that the number of reserved seats (cell K1) = VoseSimTable(Q2,Q12,,37). Running the simulations (remember to set the number of simulations to 11 in the ModelRisk Simulation Settings window) provides the results shown in Figure 7.7 (graphing the means and standard deviations from the SimTable results).

The highest mean revenue (shaded) occurs not at 37 reserved seats but at 34. We also see another important finding: The standard deviation is lower when we use Strategy 2: mean control. (In fact, the standard deviation rises continuously as we reserve more seats for Y class passengers.)

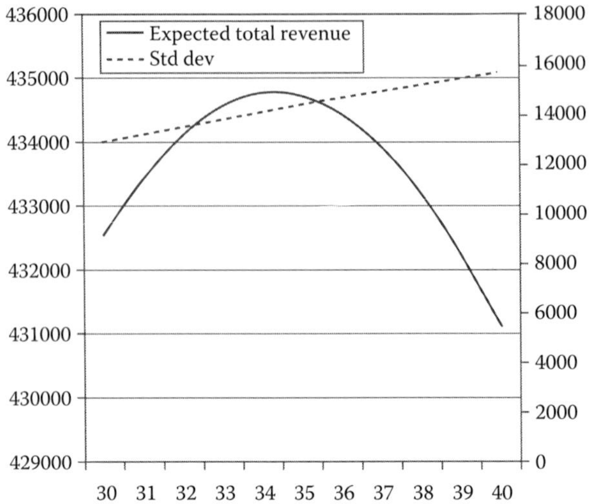

FIGURE 7.7
Graph of simulation table results.

So, the additional revenue that Littlewood's Rule provides comes at a price: increased risk. In the present case, an airline with thousands of flights per day will not care much about this increased risk since it will cancel out across these many flights (Central Limit Theory).* The expected revenues would be the more important measure for choosing a strategy and that is probably why Littlewood's Rule is stated in terms of the "expected revenue."

In fact, the application of Littlewood's Rule would require the expected Y class revenue to equal $100 (the price of a Q class seat). Since the Y class price is three times the Q class price, this is equivalent to finding where the probability of getting an additional Y class customer drops to 33.3%. In other words, we should increase the seats reserved for first class passengers until the probability of getting more first class travelers drops to 33.3% (thereby equating the expected revenue from an additional first class passenger with the economy class revenue). Using the NORMINV(0.667,30,10) function in Excel, this is found to be 34.4. Our spreadsheet model permits us to discover Littlewood's Rule as the revenue maximizing strategy.† It also shows us that maximizing revenue may not always be the best strategy since it can come at the expense of exposing us to higher risk.

* In other contexts—for example, a portfolio choice model—measuring expected return *and risk* will be equally important to the investor.

† Apparently, Boyd's use of 37 reserved seats for the application of Littlewood's Rule was an error.

7.4 An Optimization Solution to the Airline Pricing Problem

Optimization refers to finding the best solution to a problem. When a quantitative optimization is discussed, there are several important components to define when the model is set up. First, Decision Variables are quantities or values over which we have control. For example, when optimizing the allocation of money among possible investment options, the decision variables are the percentage of the portfolio assigned to each type of investment. Second, Constraints are restrictions that "constrain" the values taken on by the decision variables. In a simple portfolio example, one possible constraint is that the sum of the allocations must be equal to 100%. Third, the Objective Function of the problem is the value or quantity that we are trying to optimize—for example, the maximum return from our investment portfolio. In other situations, we might be trying to minimize a value, such as trying to minimize the risk of our investment portfolio. While in reality we often have more than one value we are trying to optimize, it is fairly difficult technically to optimize multiple objective functions. So, in practice most optimization problems are defined as a set of constrained decision variables that we are trying to manipulate so as to maximize (or minimize) a single objective function.

There are many techniques available for solving optimization problems. We examine here a small subset of possible techniques. Trial and error (such as through a simulation table as shown before) is one of the most basic optimization methods and often works very well if there are a relatively small number of decision variable values to use. However, when there are many possible values for each decision variable (as with a continuous decision variable) or too many potential combinations of decision variables (as with several decision variables, each with a number of possible values), then manual trial and error quickly becomes impractical or even infeasible. For example, if we had a fairly simply model with only 10 binary decision variables, where each could take on two possible values (zero or one), there are 2^{10} (1,024) possible combinations. Depending on the specific problem, we might be able to try all 1,024 combinations to determine the best values for the decision variables.

It is easy to see that, for realistic problems with many more decision variables, the number of possible outcomes can become astronomically large. To illustrate this, consider a problem with 50 binary decision variables (which is still smaller than many real-world scenarios). There are now 2^{50} (1,125,899,906,842,620) possible combinations. If we could check 1 million combinations per second (which is unreasonably fast for a spreadsheet model), it would still take over 35 years to check every combination in order to determine an optimal solution. Fortunately, there are many optimization software programs available that can automate and speed up the process of solving optimization problems.

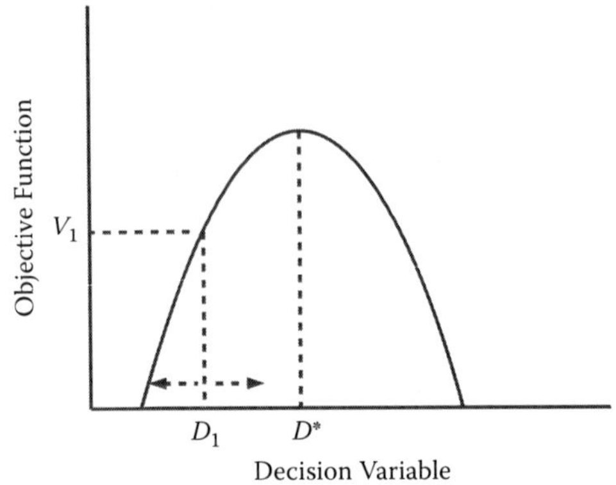

FIGURE 7.8
A simple objective function.

To understand conceptually how an optimization solver might work, consider a simple Objective Function and a single Decision Variable, as shown in Figure 7.8. Suppose that a value for the decision variable, such as D1, is put into a model and produces the result for the objective function, V1. Then, slightly higher and lower values than D1 (as indicated by the arrows) are tried, and the decision variable is moved in the direction (in this case, increasing D) that produces higher values for the objective function. If a value of D > D* is tried, then the objective function will increase by reducing the value of D. Eventually, after trying a number of different values of D, D* (or a value very close to D*) will be found to solve this optimization problem. This technique, called a "hill climbing" technique, is straightforward, provided that the objective function is well behaved as in Figure 7.8.

However, what if the objective function takes the form shown in Figure 7.9? Depending on the starting value, a simple hill-climbing method could easily decide that D1 or D3 is the optimal decision, when D2 is really superior. When there are multiple local maxima in the objective function and/or discontinuous segments (such as would be produced when there are IF statements in the spreadsheet), a smarter procedure is required. Such smarter optimization methods generally employ techniques aimed at finding the true global optimum, even in the face of such features.

These techniques differ in different optimization programs, but when spreadsheet-based models such as those with which we have been working are considered, the optimization methods generally attempt to mimic biology in finding a solution. Such evolutionary or *genetic algorithms* employ random jumps in the decision variable (analogous to mutations) to prevent premature convergence to a local, and often suboptimal, solution. By ensuring that all potential regions of the decision variable are explored, these techniques can be very effective at finding the global optimum, even when the objective function is not well behaved.

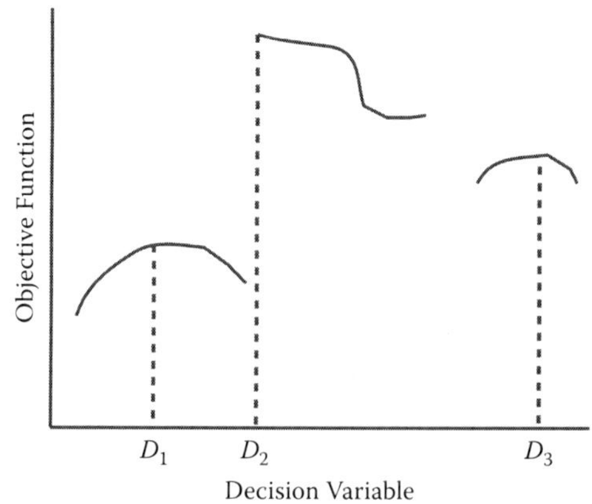

FIGURE 7.9
A more complex objective function.

Excel comes with a built-in optimization procedure called "Solver." It is found under the Data ribbon of the Excel toolbar.* The basic Solver that comes standard with Excel is limited in the type of problems that it can solve; in particular, it is not reliable when there are conditional statements (such as IF statements) in the model. Solver also frequently can get stuck in a local optimum. For example, Solver cannot handle the IF statements that are contained in our airfare pricing model and cannot accommodate Monte Carlo simulation.

Fortunately, better optimization tools are available. We will use OptQuest, which is an optimization tool included with ModelRisk that is designed to be particularly well suited for solving Monte Carlo simulation-based risk analysis models. OptQuest imposes no restrictions on the shape of the objective function, so it can be used in our airfare pricing problem. OptQuest not only will be able to find the global maximum, but it also can do so while conducting a simulation at the same time. To get a basic understanding of how OptQuest works, we will first ignore uncertainty in this section and then consider optimization under uncertainty in Section 4.5. OptQuest is capable of solving either deterministic or stochastic (i.e., with or without Monte Carlo simulation) optimization.

To conduct a deterministic optimization for the airline pricing problem, we must start by replacing the uncertain demand (probability distributions) for Q and Y class seats with a certain or known demand. The logical choice is to use the means of the distributions (90 and 30, respectively). The spreadsheet Airline pricing7c.xlsx contains this adjustment. The solution to this problem should be obvious; it is mean control (Strategy 2), where 30 seats are reserved

* If you do not see it, it is an add-in that needs to be activated. You can activate it in the same way that Data Analysis was activated in Chapter 5.

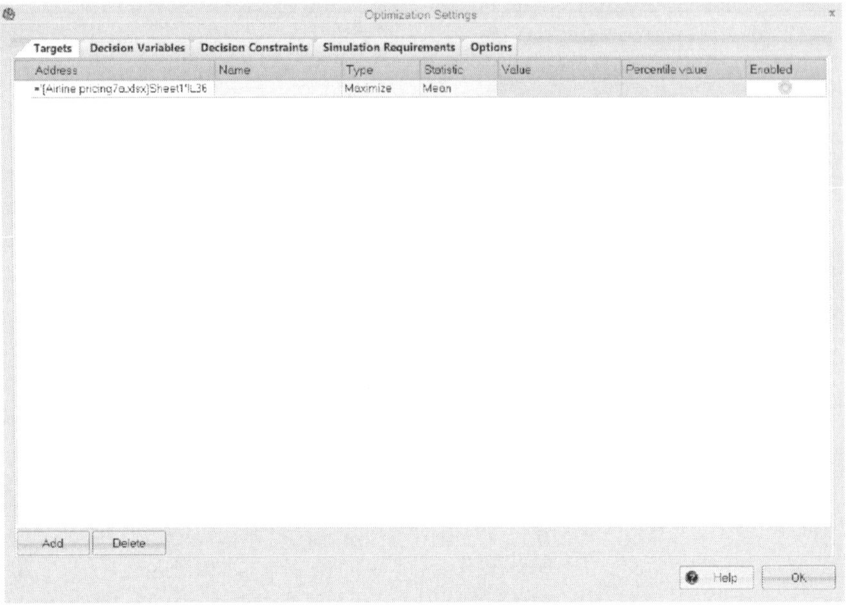

FIGURE 7.10
OptQuest target window.

for the known Y class passengers and 70 are sold to the Q class passengers. We use this simplified version of the model to show how OptQuest works and to verify that it correctly solves this problem.

Click the OptQuest button on the ModelRisk tool bar and choose Optimization Settings. This launches a window with five tabs along the top: Targets, Decision Variables, Decision Constraints, Simulation Requirements, and Options. The Targets tab is where you specify an objective to optimize. On the Targets tab, click "Add" and choose cell L36; change the Type to "Maximize" and the Statistic to "Mean," as shown in Figure 7.10. Notice that there are a number of statistics available that can be maximized or minimized, including specified percentiles of the distribution of a target cell. In this deterministic example, most of the available statistics are not relevant because our target cell is deterministic and will only change if the number of each class of seats sold is changed.

The next step is to specify one or more Decision Variables. Choose cell K1, name it "reserved seats," and set the Mode to Discrete, with a Lower bound of 28, an upper bound of 38, steps of 1, and a Base value of 28, as shown in Figure 7.11.

Decision variables (multiple variables are permitted) can be discrete or continuous. It is recommended that you start with discrete settings, even

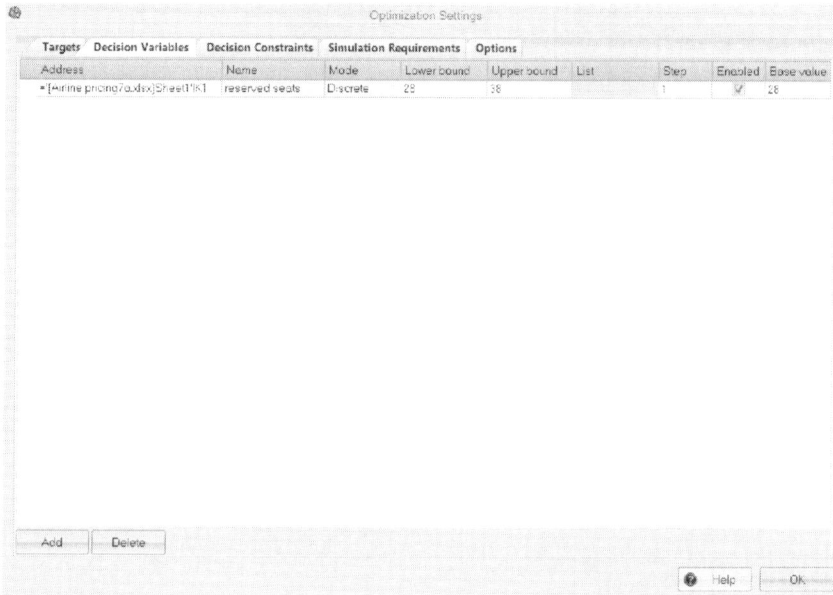

FIGURE 7.11
OptQuest decision window.

if your decision variable is in reality continuous. It is also best to choose a range with fairly large steps between discrete values to get a sense of how the target cell varies throughout the range of the decision variable. Once you have an idea of where the optimum solution may lie, you can narrow the range, decrease the step sizes or specify the decision variable as continuous, and repeat the optimization. We do not recommend changing the Mode to continuous until you have a good idea of where the solution lies. This is because a continuous random variable can essentially take on an infinite number of values, so this can be quite time consuming for OptQuest. Thus, we suggest using an iterative procedure with OptQuest: Start with large discrete steps, narrow the range and step sizes, and eventually consider continuous mode over a fairly small range if that level of precision is required.

In our current problem, there is no need to perform these iterative steps since we have a good idea where the solution lies and the decision variable (number of seats reserved for Y class passengers) is a discrete value. The next step, which deals with Decision Constraints, is optional. No constraints are necessary in this model. A common constraint would be to constrain the decision variables to be non-negative (unnecessary if the range is a bounded set between two positive values like we have here). We could constrain the

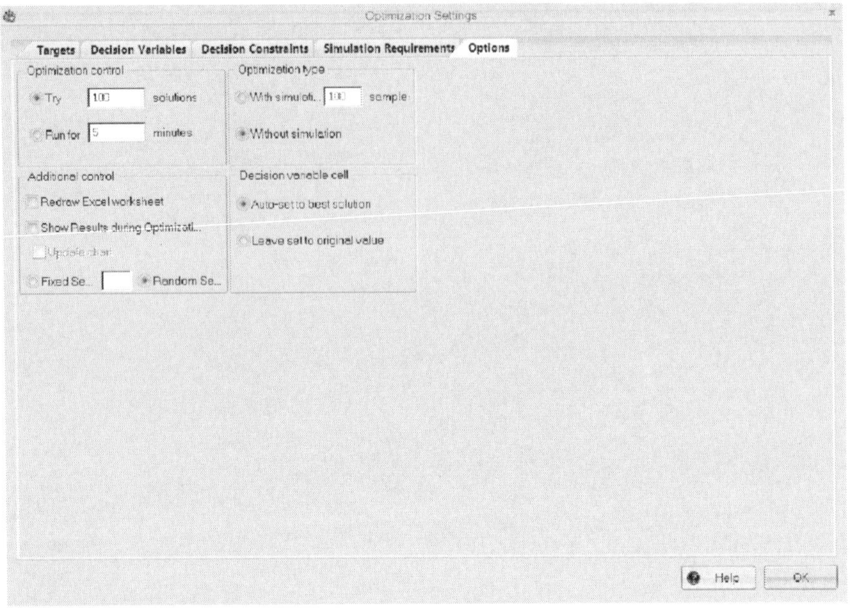

FIGURE 7.12
OptQuest settings window—no uncertainty.

total seats sold to be less than or equal to 100, but that is not necessary since we have already constructed the model to ensure that we do not exceed the available capacity of the plane.*

Simulation Requirements is also an optional setting; we will skip it now and return to it in the next section. Options will be considered further in the next section, but for now we only need to set the Optimization type to "Without simulation," as shown in Figure 7.12, to indicate that this is an optimization without Monte Carlo simulation. (This makes the other settings irrelevant, so we will address them as well in the next section.) Click "OK" and then choose Run Optimization from the OptQuest menu button and obtain the results shown in Figure 7.13.

We see that OptQuest was indeed able to find the correct solution to reserve 30 seats for Y class customers (yielding total revenue of $480,000). If you click on the Solution Analysis tab (rather than the Best Solution tab),

* The capability to set a constraint that seats sold not be greater than the plane's capacity does mean that we could have constructed the model differently, leaving the OptQuest constraint to handle this. We do not recommend this, however. The model would not operate correctly, unless you were running OptQuest, if you omitted the plane's capacity from the model. It is advisable to save the constraints for relationships that are not easy to accommodate in the original model.

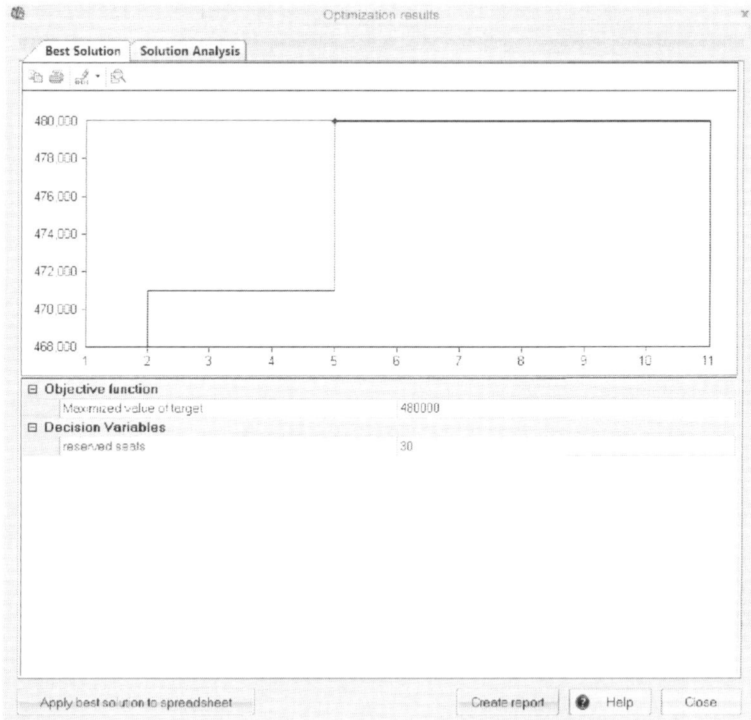

FIGURE 7.13
Deterministic simulation results.

you get Figure 7.14, which shows the results and ranking for all the values of the decision variable that were tried. If there are constraints, this view can also be used to show solutions that violated the constraints (infeasible solutions)—useful information if there is the possibility of relaxing constraints, for example, by devoting more resources to a project. If many values of the decision variable are considered, this view can be adjusted to show the best choices. In a complex optimization simulation, this is useful in order to see whether a number of choices for the decision variables provide similar results for the objective.* Finally, there is the option to paste the solution into the spreadsheet and to create a report of the optimization.

* In complex problems, especially with continuous decision variables, OptQuest may not be able to identify "the absolute best" solution, so the question is whether OptQuest has found a solution sufficiently close to the best one. One way to tell is whether the top several solutions are similar in the values for the decision variables and the value of the target. If they are not, then it suggests that OptQuest needs to be run for a longer time since there appear to be a variety of different ways to obtain similar solutions to the problem.

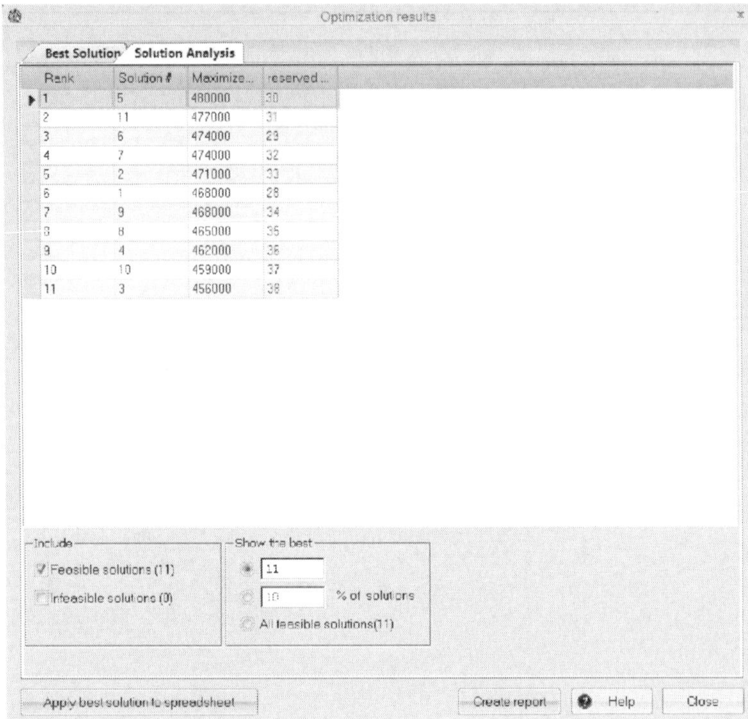

FIGURE 7.14
Deterministic solution analysis.

7.5 Optimization with Simulation

When demand for seats is uncertain (as in our original airline pricing model), the optimal deterministic strategy of mean control is no longer best. If 30 seats are reserved for Y class customers, there are times that seats will go unfilled that could have been sold to a Q class customer (an opportunity cost), and there are times when there may have been enough Y class demand to fill some of the Q class seats with a higher paying customer. Given the large difference in prices for the two classes, it seems reasonable that more seats should be reserved for potential Y class travelers when there is uncertain demand. In fact, our simulation table in Section 7.2 already confirms this to be the case.

To optimize total revenue in the face of uncertain demand, we return to our initial model and set up OptQuest to run in simulation mode (Airfare pricing7d.xlsx). The only setting that changes is under the Options tab of the Optimization Settings window, as shown in Figure 7.15.

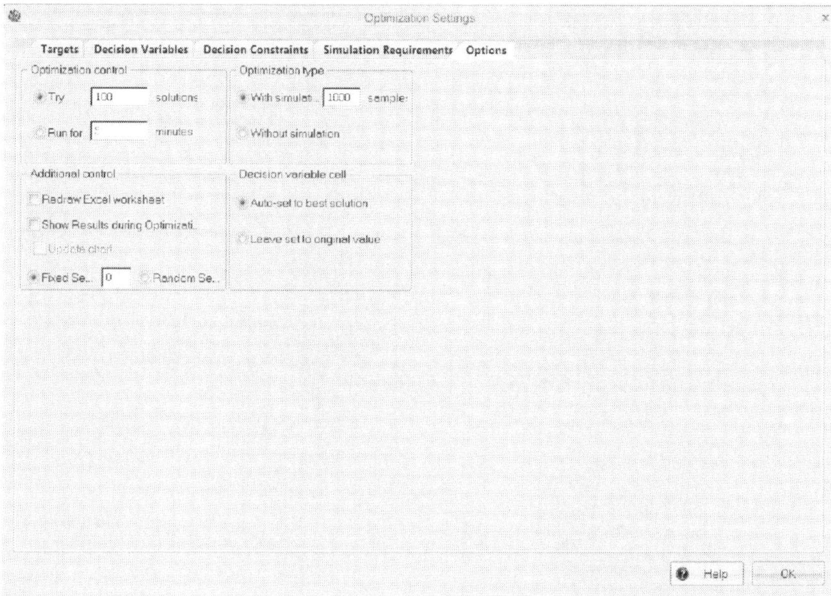

FIGURE 7.15
Stochastic optimization settings.

The optimization type has been changed from "Without simulation" to "With simulation" and the default of 100 samples has been changed to 1,000.* The number of samples refers to the number of Monte Carlo simulations (i.e., iterations) to be run for each value of the decision variable. It is usually set to a lower number than when a regular single simulation is run due to the relatively large number of simulations that OptQuest will run. However, setting the samples too low will make it difficult to determine, with much confidence, which solutions are truly better. It is important to change the radio button under Additional Control from Random Seed to Fixed Seed and to choose an initial random seed value (zero will do). This ensures that each set of 1,000 simulations will use the same set of random demand levels. If a random seed were used, then results from one value of the decision variable to another could vary due to differences in the particular random demand levels rather than differences in the decision variable.

The Optimization Control section is not important right now, but it becomes important with more complex models. When the number of combinations of values for the decision variables gets large, then OptQuest could run for a very long time indeed. It is generally better to stop it after a reasonable period (e.g., 5 minutes) and examine the solutions to see if it looks

* In our final optimization (when we get very close to the optimum), we may even change this to 10,000.

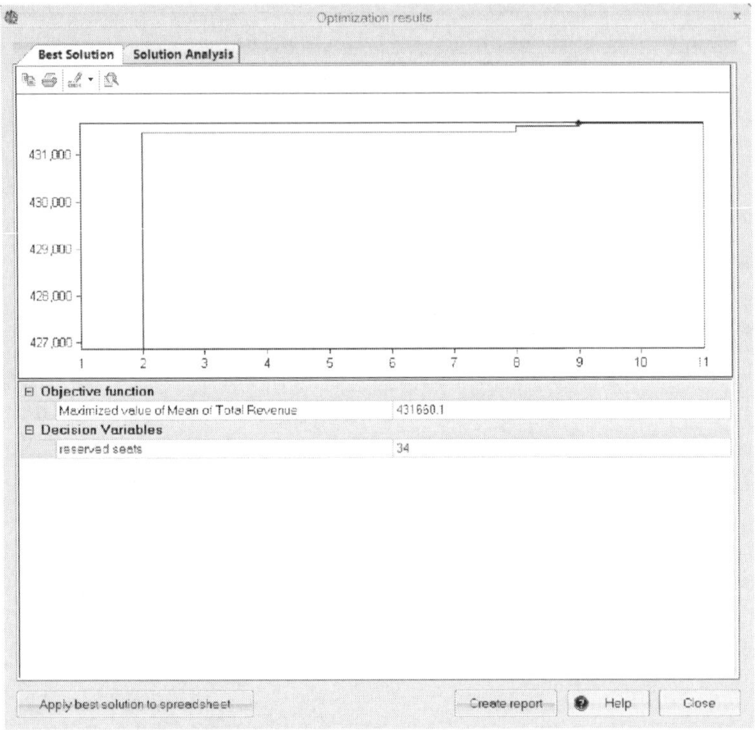

FIGURE 7.16
Stochastic optimization results.

like OptQuest has converged on a solution or not. When convergence on a solution is not apparent, that is a signal to try more possible combinations of decision variables using OptQuest. Typically, large optimization models will take hours to run before an acceptably accurate solution is reached.

After running the optimization, the results window shown in Figure 7.16 appears. OptQuest confirms what we found in Section 7.2: Reserving 34 seats (Littlewood's Rule) is the best strategy. The Solution Analysis in Figure 7.17 provides results equivalent to the simulation table we discussed in Section 7.2.

In fact, it is important to realize that a simulation table often provides the necessary information for decision making. The trial and error logic is perhaps more transparent with the simulation table than with OptQuest, the latter being something of a "black box." For this reason, very simple optimization problems may be better addressed using simulation tables. The need for optimization becomes increasingly apparent when the number of decision variables increases (Section 7.6) or if we complicate the decision problem by imposing requirements (Section 7.7).

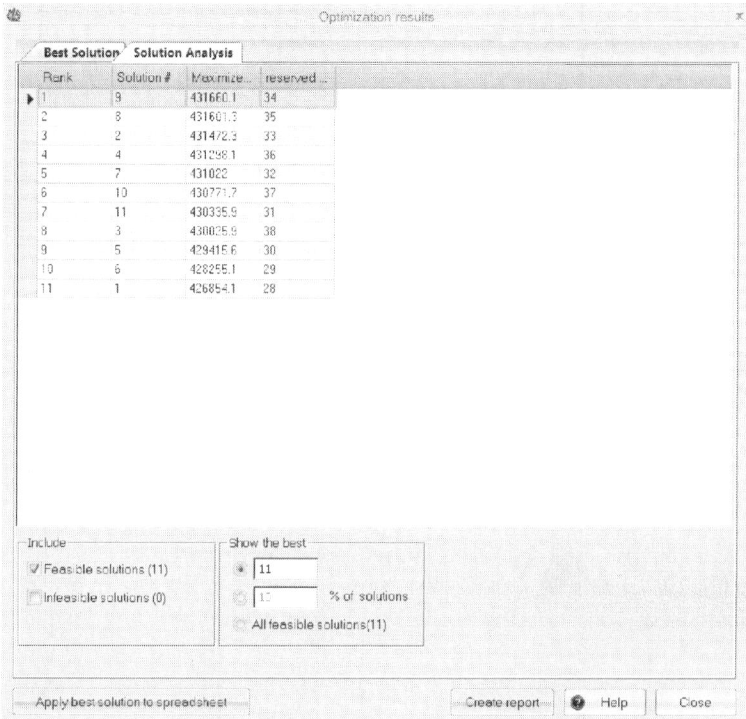

FIGURE 7.17
Stochastic Solution Analysis.

7.6 Optimization with Multiple Decision Variables

The real airline pricing problem is much more complicated in several ways. The problem we have thus far discussed considers only one route at a time in a complex network of routes. Further, the airline could set different prices for the two fare classes (not necessarily the \$100 and \$300 prices) and could establish additional fare classes. In this section, we tackle the complications associated with setting prices as well as reserved seats for Y class passengers. It is not sufficient simply to designate prices as decision variables; to do so, we would discover that the higher the prices are, the higher the revenue is. We must alter the structure of the model so that raising the prices (for either class) will negatively impact the demand for seats.

Assume that every \$10 increase in the Q class price will lead to a decrease of 20 in the expected demand. Also assume that each \$10 increase in the

	A	B	C	D	E	F
1	Seats	100		price	Demand	mean
2			Y class	300	=VoseNormalObject(F2,10,VoseXBounds(0,))	=180-0.5*D2
3			Q class	100	=VoseNormalObject(F3,40,VoseXBounds(0,))	=290-2*D3

FIGURE 7.18
Modeling prices.

Y class price leads to a decrease of 0.5 in the expected Y class demand.* Assume that both demand functions are linear and that the standard deviations of demand are unaffected by price. The required changes to the spreadsheet appear (Airline pricing7e.xlsx) in Figure 7.18.

The mean demands (F2:F3) are now expressed as a function of the prices (D2:D3), and these cells are used in the Normal distribution Objects (E2:E3).† Note that the truncation of the demand distributions to prevent negative values becomes more important now. Even though we will only consider prices that would provide for a positive demand, the variability in the simulations could easily produce negative demands if we did not truncate the distributions. The last required change is on the Decision Variables setting. We need to add the two prices as two additional decision variables.

We must be a bit careful with the ranges for the prices. Given our previous assumptions, Q class prices above $145 would generate negative demand, as would Y class prices above $360. Even though we have not permitted demand to be negative, it makes no sense to consider prices in these ranges. Also, negative prices do not make sense. So, we restrict prices to be within these ranges. For our initial simulation, make the prices discrete with fairly large steps of $10; we want to get a sense of what the optimal solution might look like before trying to get more precise. We also consider reserved seats between 30 and 100 with large steps since we really do not know what the optimal solution might look like for the new problem. Figure 7.19 shows the decision variable settings.

After 100 simulations, OptQuest stops. (That was the setting in the Options tab; there are 4,440 combinations of prices and reserved seats [15 Q prices × 37 Y prices × 8 seat levels = 4,440 combinations] that OptQuest could try, so it is not done yet.‡) At this point, it reports that the best solution it has found is to reserve 70 seats with a Q class price of $140 and a Y class price of $220. Figure 7.20 shows the Solution Analysis at this point.

* These assumptions are consistent with the Y class demand being more inelastic than the Q class demand.

† The Lognormal distribution could be a better choice for the demand distribution because there is likely some skew where demand is more often lower than the mean than above it. However, we will continue to use the Normal distribution here to stay consistent with the earlier example.

‡ OptQuest is very smart, though, and does not always have to evaluate every possible scenario to find the optimum.

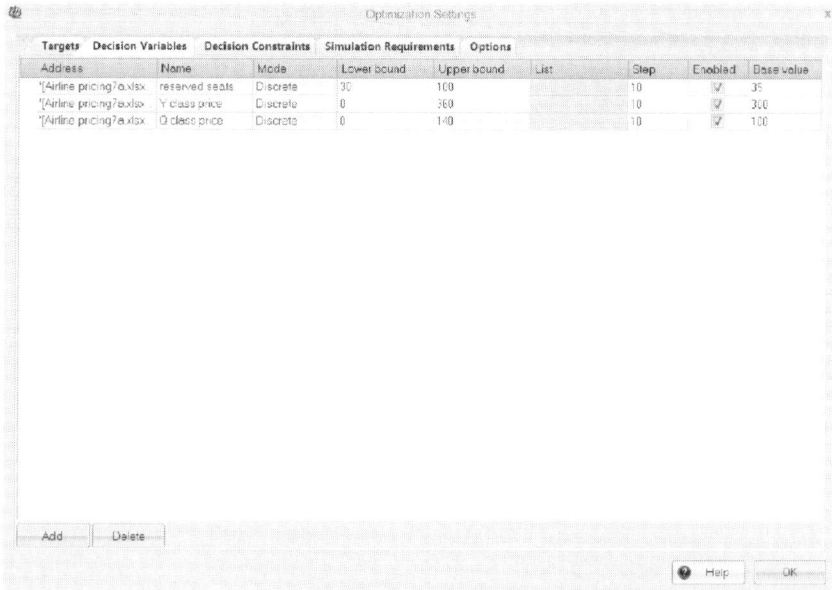

FIGURE 7.19
Setting three decision variables.

The top solutions do not vary much: Y class prices fall in the range of (210,230), Q class prices within (120,140), and reserved seats mostly at 70. We will refine the analysis by using these price ranges, permitting the reserved seats to vary between 65 and 75, and using smaller steps of one for each variable. Notice that we could set a Y class price of $160 and have a mean demand of 100, thereby (on average) filling the plane with Y class passengers. OptQuest apparently does not find that to be as profitable as charging higher Y class prices and having both classes of passengers on the plane.

With the narrower ranges for the decision variables, the optimization (after 100 solutions, now out of 4,851 possible combinations) produces the Solution Analysis shown in Figure 7.21. At this point, the best solution is to reserve 66 seats for Y class, with a price of $220 and sell Q class tickets at a price of $139. It is left for the reader to explore finding a more precise solution. (After all, price is really a continuous variable.)

7.7 Adding Requirements

Using the decision variable values determined in Section 7.6 (66 seats for Y class, and prices of $220 for Y class and $139 for Q class) and running

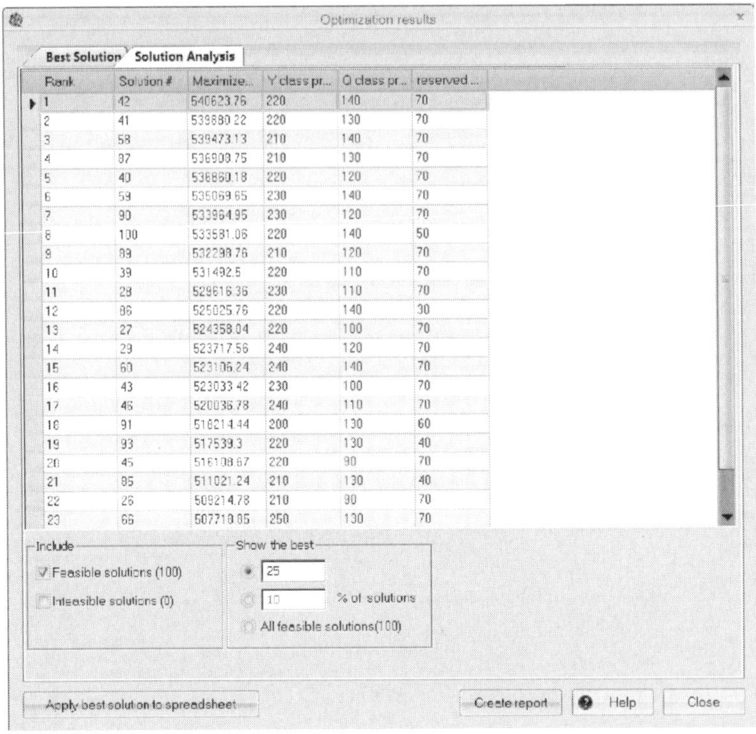

FIGURE 7.20
Solution Analysis version 1.0.

a simulation of 10,000 samples give the results for total revenue shown in Figure 7.22.

Mean total revenue is \$541,804 and a 90% confidence interval ranges from \$524,352 to \$557,823. Suppose that the standard deviation of \$10,168 is deemed to be too high (resulting in the confidence interval being too wide). Is it possible to reduce the standard deviation to \$9,000?* If so, how is that accomplished with the three decision variables and what is the cost in terms of reduced total revenue? OptQuest can answer such questions through the use of *Simulation Requirements.* Simulation Requirements are similar to Constraints in that they limit the potential solutions of an optimization problem. However, Constraints operate on decision variables while Simulation Requirements can apply to any variable or component in the model. The idea is that a feasible solution requires a particular component of the model to

* While it is not very realistic for an airline to be concerned about the standard deviation on a single route, in other contexts, reducing the standard deviation may be of considerable interest. For example, in a financial asset portfolio optimization, investors are always concerned about earning a high return, but are also interested in reducing their risk (in the financial field, often measured by the standard deviation of the returns).

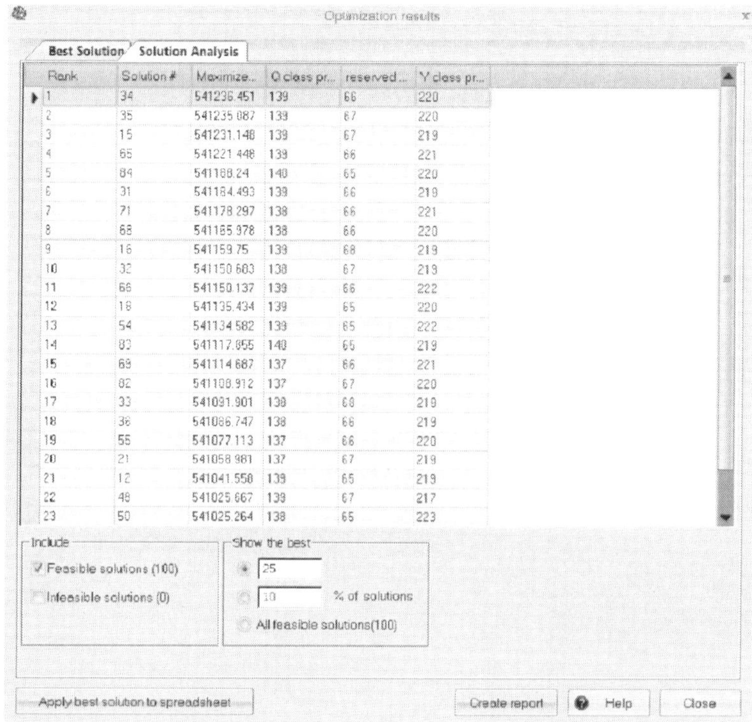

FIGURE 7.21
Solution Analysis version 2.0.

possess certain desired characteristics. In contrast, a Constraint means that decision variables must obey particular relationships.

To enter a requirement, use the Simulation Requirements tab of the OptQuest settings, as shown in Figure 7.23. A wide variety of parameter requirements is possible (including means, standard deviations, percentiles, values, etc.). In this example, we would like to limit the standard deviation of total revenue to 9,000 or less. Since we do not know if this is feasible or how it might be accomplished, we also change the decision variables back to wide ranges and large steps. The solution analysis, after 5 minutes, appears in Figure 7.24.

While we do not show a very precise answer at this point, we do know that a solution to this problem is feasible (although OptQuest only found nine feasible solutions in 5 minutes). The best of these produced mean revenues of $535,272 (a reduction of $6,532 from the solution before adding the requirement). Reducing the risk appears to require a greater divergence in prices between fare classes and a reduction in the number of reserved seats. Again, characterizing the optimal solution more precisely is left for readers to explore.

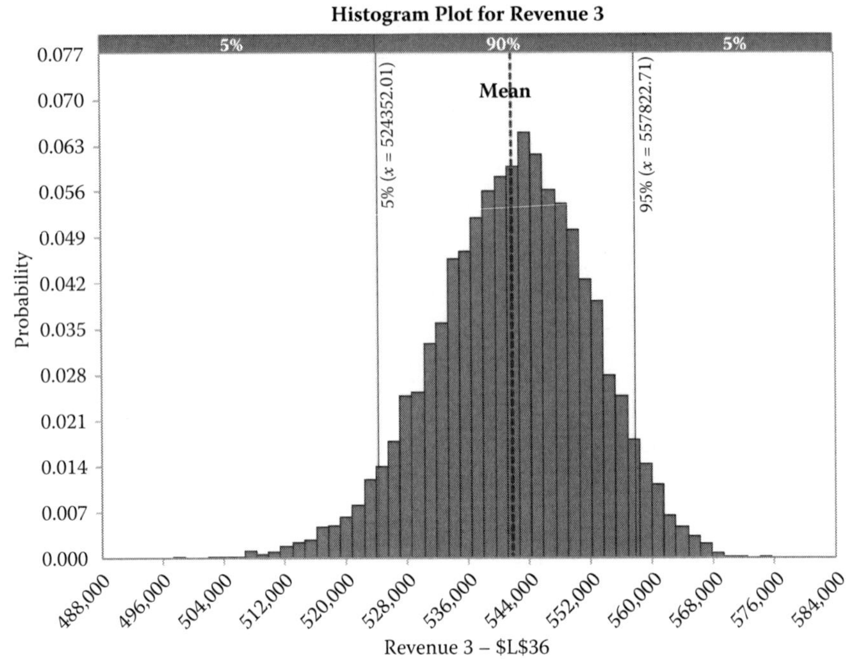

FIGURE 7.22
Distribution of the optimal solution.

One further step would be to vary the requirement—for example, to see how restricting the standard deviation of total revenue to be less than $2000, $4000, $6000, and $8000 would impact the optimal solution. In the case of an investment portfolio, if the requirement is set to restrict the probability of a loss to 5, 10, 15, 20%, etc., then each of these restrictions would entail a different optimal solution and a different level of return. These risk–return trade-offs are referred to as an *Efficient Frontier.* The term is widely used in finance, but it can refer to any relationship that shows the optimal value of an objective function as a function of various levels of a simulation requirement. Construction of an efficient frontier entails running the optimization at a number of values for the simulation requirement, collecting the results (objective function values and requirement levels), and then displaying the frontier (e.g., the maximum return for each level of risk).*

* Crystal Ball has an automated capability to produce an efficient frontier. At present, other Monte Carlo spreadsheet programs, including ModelRisk, require this construction manually. Crystal Ball refers to this as a "variable requirement." One reason for employing a variable requirement to build an efficient frontier is that the decision maker may be unsure what level of risk he or she is willing to tolerate, so an estimate of the trade-off between risk and return would be most useful.

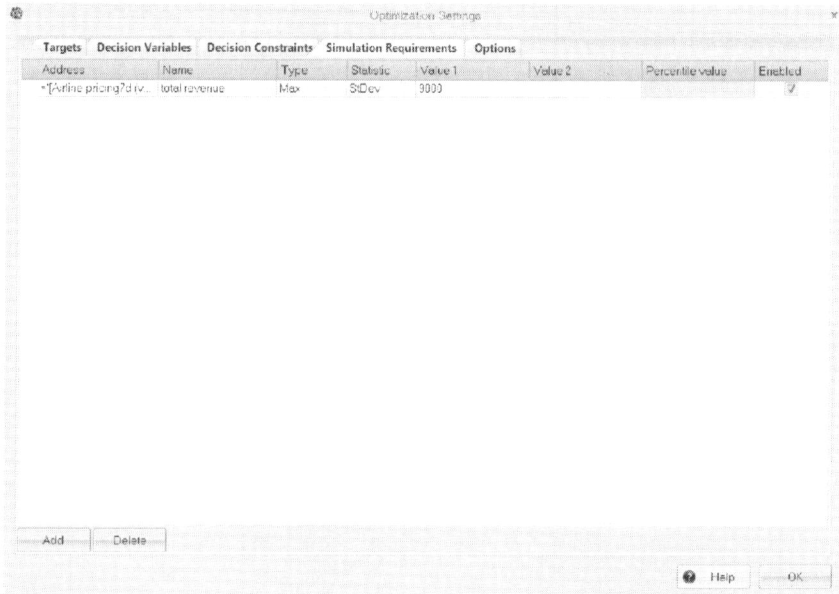

FIGURE 7.23
Adding a requirement.

7.8 Presenting Results for Decision Making

The primary purpose of using Monte Carlo simulation is decision support. To be useful, this requires that the assumptions and results of an analysis be clearly presented. We offer a few guidelines here, based on our experience with the use of Monte Carlo simulation models in management.

It is best to avoid technical jargon and complex mathematical language whenever possible. This does not mean that things should be oversimplified; however, it does mean that they should not be any more complex than required in order to capture the important elements of a decision problem. As Edward Tufte has stated about information design: "What is to be sought in designs for the display of information is the clear portrayal of complexity. Not the complication of the simple; rather, the task of the designer is to give visual access to the subtle and the difficult—that is, the revelation of the complex."*

* Tufte, E. 2001. *The Visual Display of Quantitative Information*, 2nd ed., 9, 191. Cheshire, CT: Graphics Press.

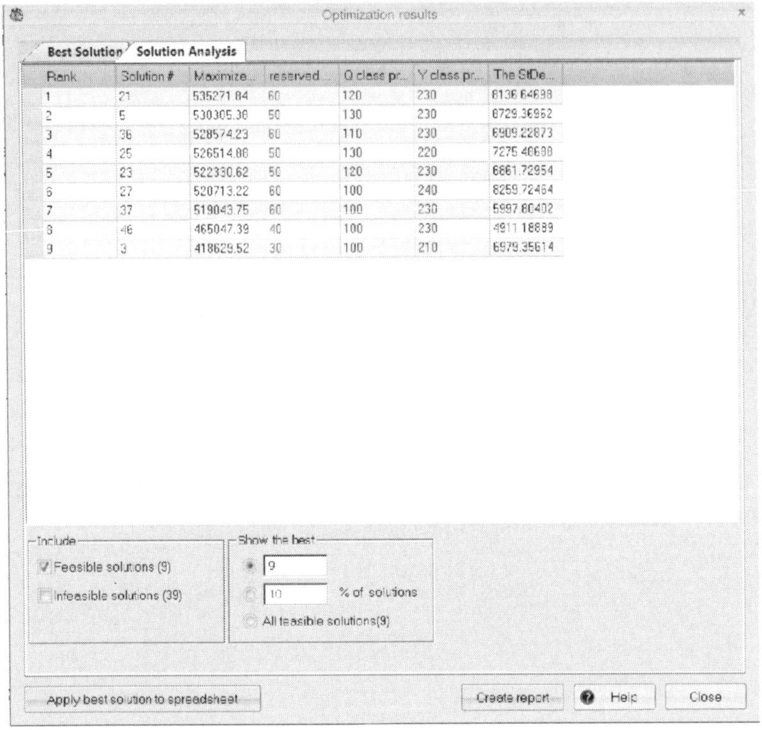

FIGURE 7.24
Initial Solution Analysis with a requirement.

Monte Carlo simulation can be useful for examining complex future scenarios where uncertainty is inherently important. An analysis should convey that multiple scenarios are possible; each trial of the simulation represents a different possible future.* The average (or median, or most likely) of these scenarios is almost always of interest, but so is the variability of possible scenarios. Modeling averages without uncertainty does not require Monte Carlo simulation, but modeling variability and uncertainty is greatly enhanced through simulation.

Consider a simple conceptual example: A decision maker is considering building a privately financed and operated bridge. The potential costs and benefits of constructing the bridge must be estimated, as well as the impacts

* In fact, in Monte Carlo simulation, it is important that every iteration be a possible future scenario.

of setting various prices (tolls) for the use of the bridge. Estimating the expected profitability of the project will be important, as well as the probability that the project will not be profitable. Risk analysis models may differ in how they measure these two features—risk and return—but usually both need to be understood.

Presentation of model results (whether an optimization model or an exploration of possible future scenarios without explicit optimization) always requires a clear description of the important assumptions of the model. It is a good idea to describe how alternative important assumptions might affect the results. For example, an assumption will need to be made concerning how changing the toll will impact the willingness of drivers to use the bridge. A good model should convey how variations in this assumption might impact the profitability, including the risk around the profitability, of the bridge project.

Monte Carlo simulation models provide an (over)abundance of quantitative information. It is not a good idea to overwhelm decision makers, but some outputs are essential. First, some measure of the expected outcome (mean, median, most likely scenario, etc.) should always be provided. Second, some measure of its variability (frequently providing P10 and P90 or the probability of particularly important outcomes, such as the NPV > 0) should be included.* Graphical displays are often the best way to convey results. To illustrate the future scenarios for the NPV of a bridge project, some display of the distribution of results should be employed. This may be a histogram, a box plot, or a cumulative distribution. If the outputs of interest involve a sequence of values over time (for example, the cash flows associated with the bridge), then trend charts are a good way to convey the future scenarios. A useful method to visualize the importance of different uncertain inputs to a model is through a Tornado sensitivity chart.

Projects involving risk typically result in trade-offs between risk and return. It is the nature of these trade-offs that a modeler should seek to describe. Sometimes the decisions will be easy; for example, if setting a relatively low toll on the bridge enhances its expected profitability and also reduces its risk, then lower tolls are likely to be preferred. However, often the decisions will not be so obvious. Higher returns frequently come at the expense of higher levels of risk. It is precisely these types of decisions where a good simulation model can be most useful.

* P10 refers to the 10th percentile of the distribution (i.e., there is a 10% probability that the outcome will be less than or equal to this value). Percentiles or intervals such as (P5, P95) or (P10, P90) are generally much more intuitive than measures such as the standard deviation.

7.9 Stochastic Dominance*

In general, when comparing distributions of two different outcomes, it will not be obvious which one is better. It is easy to see which has a higher payoff, but rarely will a strategy that results in maximizing a payoff also be a strategy that minimizes risk. There is usually a trade-off involved. Simulation tables can be very useful in helping with decisions involving trade-offs because the risks and returns can be readily compared. More systematic tools have been developed, however, that are designed to quantify differences between distributions that embody different risk/return profiles.

Stochastic dominance is one such technique. ModelRisk contains a feature that allows the user to determine stochastic dominance, which enables you to compare two (or more) distributions and determine whether one is "better" than the other. When using a stochastic dominance test, the most common comparisons involve determining whether one distribution stochastically dominates another in the first degree or the second degree. First degree stochastic dominance means that one cumulative distribution is always to the right of the other.

Consider the distributions for the profitability of two alternative projects shown in Figure 7.25. It is not obvious which project is to be preferred. Project A appears to have a higher expected return, but also appears to be somewhat more variable. However, if we view the cumulative distribution functions, the choice becomes clear. Figure 7.26 shows the cumulative probability distributions. In this figure, Project A's cumulative distribution of the objective function is always greater than that for Project B. This means that there is always a greater probability that Project A will reach any given level of profitability than will Project B. Clearly any risk-averse decision maker should prefer Project A to Project B.

First degree stochastic dominance could happen, for example, if the distributions for A and B were identical, except that A had a higher mean than B. For instance, if A and B were both Normal distributions, but A had a higher mean than B, then the graphs of their cumulative distributions would look similar to Figure 7.26, and A would stochastically dominate B in the first degree. The importance of this is that any decision maker should prefer option A to option B since the payoff is always greater for any percentile.

Many other scenarios will not be so easy to compare. Consider Figures 7.27 and 7.28. The cumulative distribution for option C lies above that for option D for some values of the objective function, but not for others. In other words,

* This section is more advanced and can be skipped without loss in continuity. Stochastic dominance is not extensively used in management practice, although it is of considerable theoretical interest and is increasingly being employed in some areas. For example, see the recent book by S. Sribonchita et al. 2009. *Stochastic Dominance and Application to Finance, Risk, and Economics*. Boca Raton, FL: Chapman & Hall/CRC Press.

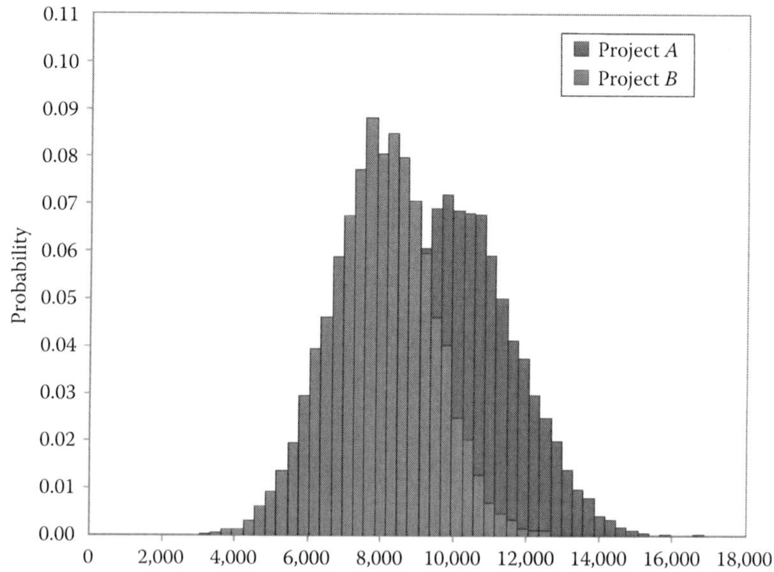

FIGURE 7.25
Density functions exhibiting first degree stochastic dominance.

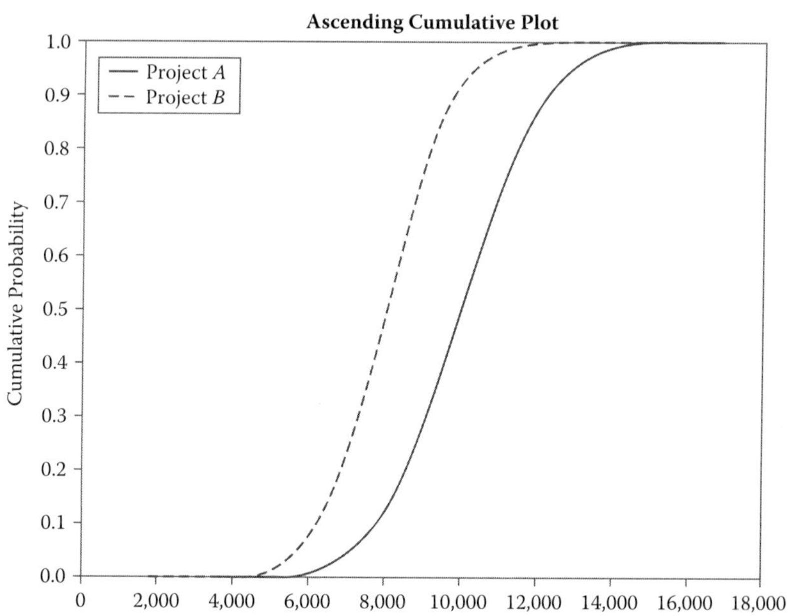

FIGURE 7.26
Distribution functions exhibiting first degree stochastic dominance.

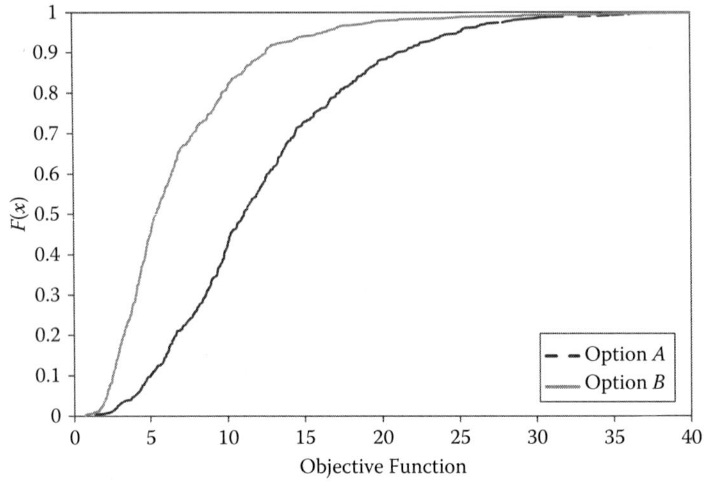

FIGURE 7.27
Another pair of probability density functions.

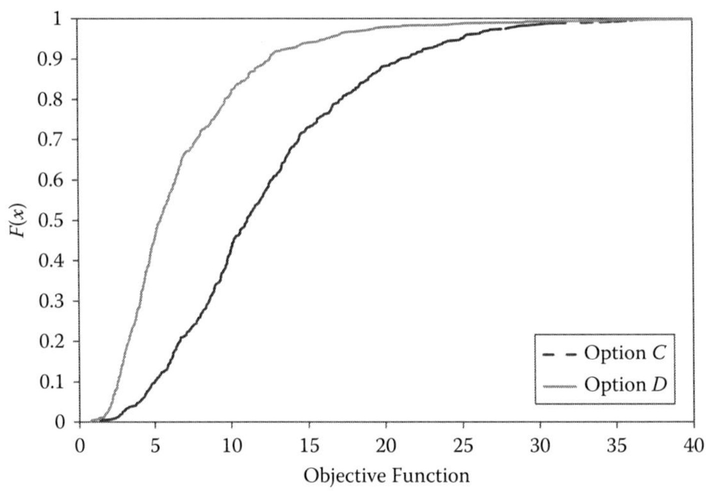

FIGURE 7.28
Second pair of cumulative distribution functions.

the situation is not as clear as it was in the previous example. In fact, comparing such options depends on the decision maker's preferences—particularly, his or her attitude toward risk. If risk is well tolerated, then the decision maker may prefer option D, which entails more risk (the probability of low payoffs is higher) in exchange for greater potential returns. (The cumulative probability for C becomes greater once high enough return levels are

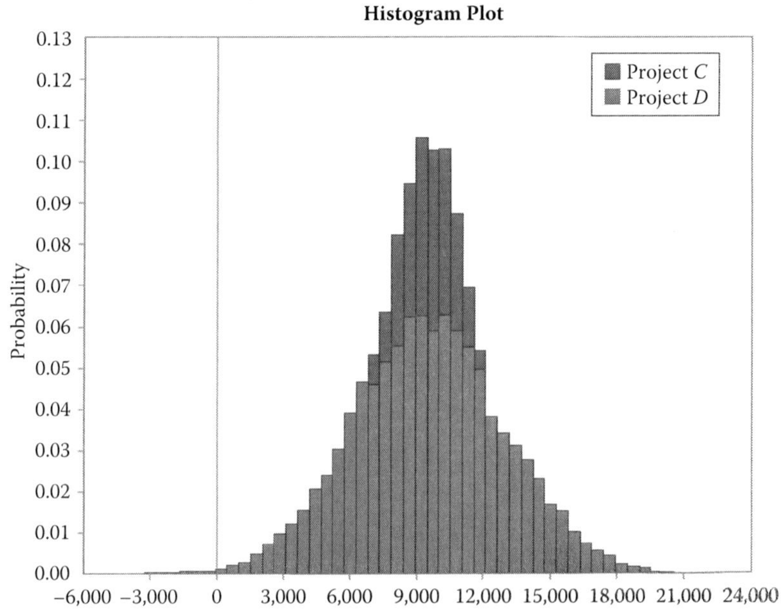

FIGURE 7.29
Density functions exhibiting second degree stochastic dominance.

reached, meaning that D has a greater probability of attaining yet higher levels.) Conversely, a risk-averse decision maker will prefer to avoid riskier outcomes and may prefer C to D. In general, such comparisons are often ambiguous and highly dependent on the decision maker's preferences, so it is advisable to use tools such as simulation tables to compare the distributions for different decisions fully.

Sometimes a stochastic dominance test may help. A distribution dominates another in the second degree when any risk-averse decision maker would prefer one to the other. A simple example would be two distributions with the same mean, in which one has a higher standard deviation than the other, as shown in Figure 7.29. In cases like this, any risk-averse decision maker would prefer option C to option D since it has less uncertainty but the same expected payoff.

Now, can stochastic dominance be applied to our airline pricing problem? In Section 7.2, we produced a simulation table that revealed that reserving 34 seats for Y class produced the highest expected revenue, but that reserving only 30 seats produced the lowest standard deviation. Can we compare these two strategies using stochastic dominance?

Airline pricing7f.xlsx is the original spreadsheet (Airline pricing7a.xlsx) adapted to consider stochastic dominance between the two strategies: Reserve 30 seats (Revenue 2) and reserve 34 seats (Revenue 3). The two strategies (30 and 34 reserved seats) have been inserted into the Strategy 2 and

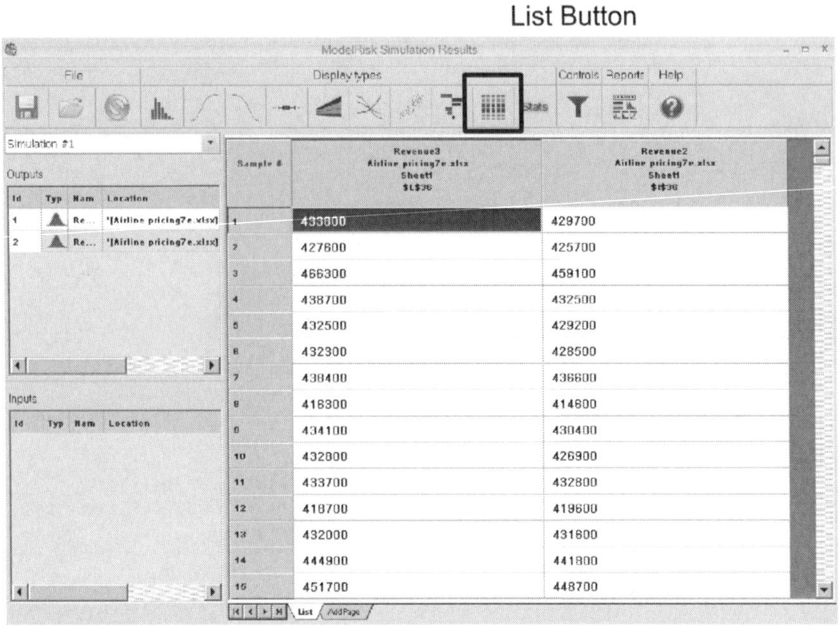

FIGURE 7.30
Extracting simulation outputs.

	W	X	Y
1	=VoseDominance(T2:U10001,T1:U1)	=VoseDominance(T2:U10001,T1:U1)	=VoseDominance(T2:U10001,T1:U1)
2	=VoseDominance(T2:U10001,T1:U1)	=VoseDominance(T2:U10001,T1:U1)	=VoseDominance(T2:U10001,T1:U1)
3	=VoseDominance(T2:U10001,T1:U1)	=VoseDominance(T2:U10001,T1:U1)	=VoseDominance(T2:U10001,T1:U1)

FIGURE 7.31
VoseDominance array functions.

3 columns, respectively. To use the stochastic dominance test, we need to compare the output distributions for the two strategies. We run the simulation and then extract the simulated values for the two outputs using the List selection, as shown in Figure 7.30, and copying and pasting all the values into columns S, T, and U of the spreadsheet.

The VoseDominance array function uses the contiguous array of the data being compared and the labels that go with these series (optional). Figure 7.31 shows the array function in the spreadsheet. The first range of cells contains the simulation outputs, while the second (optional) has the labels. In this case, the matrix displays "Inconclusive" because neither distribution stochastically dominates the other in either the first or second degree.

To see a version of our airline pricing example where the stochastic dominance test does provide some definite results, consider changing the reserved seating for Y class passengers to 40 seats, and compare this with the mean

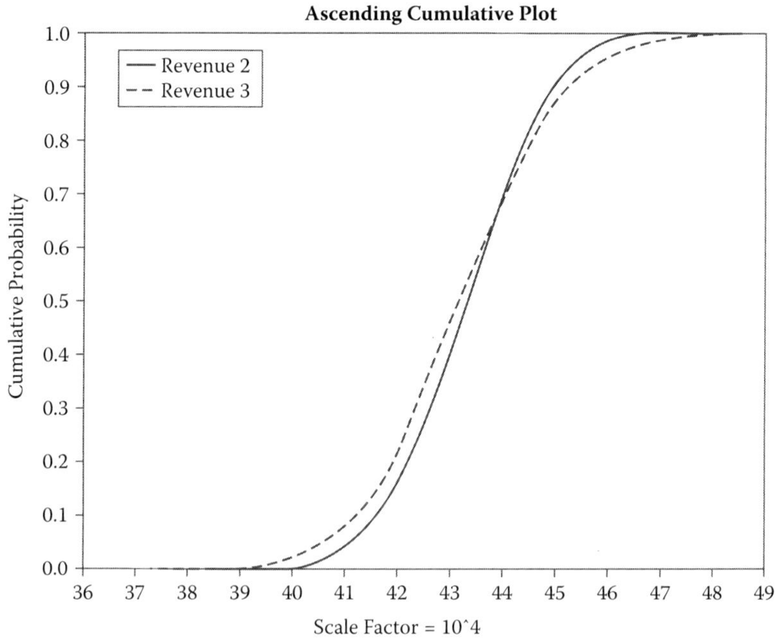

FIGURE 7.32
Cumulative distributions for 30 and 40 reserved seats.

	W	X	Y
1	Revenue2 is 2d over Revenue3	Revenue2 is 2d over Revenue3	Revenue2 is 2d over Revenue3
2	Revenue2 is 2d over Revenue3	Revenue2 is 2d over Revenue3	Revenue2 is 2d over Revenue3
3	Revenue2 is 2d over Revenue3	Revenue2 is 2d over Revenue3	Revenue2 is 2d over Revenue3

FIGURE 7.33
Stochastic dominance results for 30 and 40 reserved seats.

control (30 reserved seats) strategy. Running the simulation provides the two cumulative probability distributions shown in Figure 7.32.

Neither strategy exhibits first degree stochastic dominance, but it appears that a risk-averse decision maker might prefer Strategy 2, given that the probability of receiving all of the lower revenue levels is greater with Strategy 3 than with Strategy 2. But Strategy 3 does have higher probabilities of achieving higher revenue levels than Strategy 2. The stochastic dominance test does confirm that Strategy 2 stochastically dominates Strategy 3 in the second degree ("Strategy2 2d over Strategy3") and thus is preferred by all risk-averse decision makers, as shown in the results of the VoseDominance array function in Figure 7.33.

Stochastic dominance tests can provide a measure with which to compare the performance of different strategies where each strategy has different return and risk levels. Other techniques for making decisions with multiple objectives (some with optimization capabilities) fall under the names of Multiattribute Decision Making or Multicriteria Decision Making.* These topics are beyond the scope of this book.

7.10 Summary

There is a joke about a traveler landing in New Zealand and renting a car. The traveler knows that New Zealand was part of the British Empire but cannot recall whether they drive on the left or right side of the road. So, the traveler decides to drive in the middle, with the reasoning that, on average, it is the correct choice.

Such "optimization," using the "average" of possible outcomes and ignoring the varying scenarios, may only hit the target of disaster. Used properly, optimization can greatly assist with decision making under uncertainty. In particular, when there are multiple decision variables or when attitudes toward uncertainty can be explicitly specified, the use of optimization tools can be quite robust.

Simulation tables and graphical exploration of simulation results can always be considered as an alternative or complement to software-based optimization solutions. Usually, different potential decisions will imply different risks and returns. Occasionally, the choices will be easy, but most of the time they will involve complex trade-offs that are not always easy to understand and rank qualitatively. Analysts can support decision makers by elucidating these trade-offs. Monte Carlo simulation is a powerful means for doing so, and, under the right conditions, optimization together with Monte Carlo simulation will allow us to identify the optimal solution while acknowledging and taking into account risk and uncertainty.

CHAPTER 7
Exercises

7.1 QUALITY-ADJUSTED LIFE YEARS

Medical decisions are complex, and many affect both the length of life and the quality of life. Quality-Adjusted Life Years (QALYs) is a metric that has been developed to assist with such decisions.† The basic idea behind QALYs is that a year of life in perfect health equals one QALY. If

* A recent exploration of these techniques can be found in Lawrence, K. and G. Kleinman. 2010. *Applications in Multi-Criteria Decision Making, Data Envelopment Analysis, and Finance* (Applications of Management Science). Bingley, England: Emerald Group Publishing.

† A good introduction to QALYs and their measurement can be found in Sassi, F. 2006. Calculating QALYs, comparing QALY and DALY calculations. *Health Policy and Planning* 21:5, 402–408.

TABLE 7.1

Treatment Options

Treatment	Cost/ Treatment ($)	QALYs	Expected Lifetime (Years)	Year of Full Recovery
A	131,200	8.26	10.2	5
B	94,800	7.68	9.1	4
C	76,200	6.48	8.7	Never

a person is disabled, or in sufficient pain, a single year might be equivalent to only half of a QALY. More (less) severe impairments equate to smaller (larger) QALYs. Many medical treatments vary in terms of how they influence both the length and quality of life. For example, more invasive treatments often involve prolonging life, but at the expense of a low quality of life, at least during the period of recovery from the treatment. Treatments also vary in terms of cost. Consider the hypothetical treatments described in Table 7.1.

Assume that the discount rate is 10%. Treatment A is fairly costly and invasive and entails a long recovery period, but full recovery is expected. Treatment C is less costly and invasive, but never completely cures the patient, leaving some chronic problems. Treatment B is between the two. The expected lifetimes reflect the average remaining life for people who contract this condition.

a. Build a model to calculate cost per present value of QALYs for each treatment. Use a 15-year time horizon and round fractional years downward to the nearest integer for calculating the benefits each year (e.g., 10.2 years is 10 years; do not use partial years for the time line, but use the values in the table for calculating the QALYs).
b. Assume that the expected lifetimes are uncertain and follow Lognormal distributions (with means equal to those given in Table 7.1 and standard deviations equal to 30% of these values). The QALY data are also uncertain; they follow Lognormal distributions (with means given in Table 7.1 and standard deviations equal to 20% of these values). Build a simulation model to compare the cost per QALY (discounted). How often does each treatment minimize the cost per QALY?
c. Suppose that you have $1 billion to spend on these treatments and that there are 20,000 people that suffer from this condition. How would you allocate your budget among the three treatments to maximize the total discounted QALYs? How many people will be treated with this solution? How would you allocate your budget to maximize the number of people treated?

7.2 FUEL COSTS VERSUS LABOR COSTS

Suppose that a truck gets 11 miles per gallon if it travels at an average speed of 50 miles per hour. Fuel efficiency is reduced by 0.15 mpg for each increase in the speed of 1 mph, up to 60 mph. Fuel efficiency is reduced by 0.2 mpg for each 1 mph in the range from 60 to 70 mpg, and by 0.3 mpg for each 1 mph above 70 mph. Drivers earn $40 per hour and fuel costs $4.20 per gallon.

a. What speed minimizes the total cost of a 500-mile trip? Does it depend on the length of the trip?
b. Suppose that fuel costs are uncertain, following a Lognormal distribution (mean = $4.20/gallon, standard deviation = $1). What speed minimizes the total expected trip cost? Again, does it depend on the length of the trip?
c. With uncertain fuel costs (as specified in part b), what speed minimizes the total cost of a 500-mile trip with a 90% probability that the total cost will be less than $600?

7.3 PORTFOLIO MANAGEMENT

Exercise7-3.xlsx contains annual data for the returns on stocks, Treasury bills, and Treasury bonds from 1928 to 2010.* Imagine that you have $99,999 to invest in these three funds and that you are currently 60 years old.

a. Build a simulation model to show the size of your portfolio over the next 5 years for an equal allocation among the three funds. Base this model on two alternative estimations for the annual return over the next 5 years: (1) using the best time series model, and (2) fitting static normal distributions to the data. You should include correlations in your model. Run a simulation and compare the values of the equal-allocation portfolio.
b. Find the portfolio allocation that maximizes your final value in 5 years (for each type of return estimation). Provide 90% confidence intervals for your returns. Suppose that the investor wishes to maximize the return, but ensure that the probability of losing money is no greater than 10%. What is the optimal portfolio?
c. Repeat part (b) for a 50-year-old investor planning for a 15-year investment period.
d. Many theories in finance are based on assuming that returns are normally distributed. Comment on how that assumption affects the results in this problem.

7.4 OPTIMAL INVENTORY MANAGEMENT

The simplest inventory model is the economic order quantity (EOQ) model, which assumes that demand is known with certainty, there is a constant lead time between order and receipt of inventory, there is always a fixed order quantity, and there are constant holding and ordering costs per unit. More realistic inventory models are often based on this basic EOQ model as a starting point. A continuous

* The Treasury bill rate is a 3-month rate and the Treasury bond is a constant maturity 10-year bond. Stock returns are based on the S&P 500.

review inventory system involves constant monitoring of the inventory on hand (and already ordered), and placing another order whenever the *inventory position* (current inventory + orders not yet received) falls below a level R, the *reorder point*. The general result in this model is that the optimal order quantity is found from Equation 7.1:

$$Q^* = \sqrt{\frac{2D(OC)}{HC}} \tag{7.1}$$

where D is the annual demand, OC is the ordering cost per order, and HC is the holding cost per unit per year. Optimal inventory policy is a selection of the order quantity, Q^*, and the reorder point, R.

a. Consider a hot tub retailer with average annual demand for 620 tubs. Ordering is expensive, at $500 per order, as are inventory holding costs ($500 per tub per year). Assume that each order takes 3 weeks from the time it is ordered until it is delivered. If an order is received and there is no stock on hand, the customer is offered a rebate of $300 for having to wait for the tub. Build a model to show 52 weeks of evolution for this business. Find the optimal reorder point and order quantity for a year that minimize the total cost. (Remember that the reorder point and order quantity are single choices and do not change over time in this model.) Given this optimal policy, what is the probability of not being able to make a sale that is demanded?
b. Suppose that demand is uncertain and follows a Poisson process with the average rate given previously. Also assume that the delivery time (the time between ordering and receiving an order) is also uncertain: There is a 50% chance that it will take 3 weeks, a 25% chance that it will take only 2 weeks, and a 25% chance that it will take 4 weeks. Build a simulation model for 52 weeks and find the optimal order quantity and reorder point for this model. With the optimal policy, what is the probability of not being able to fulfill a retail demand when it arrives?

7.5 PROJECT SELECTION

A mobile phone company must decide the types of R&D in which to invest. Suppose that it has seven projects available, as described in Table 7.2. Assume that the potential profits (a total) will be earned over a 4-year period (equal amounts in each year). Assume that the discount rate is 9%.

a. If all the ranges in Table 7.2 are replaced by their midpoints (and are known with certainty) and you have $75 million available for R&D (total over the next 8 years), in which projects should you invest to maximize your NPV? What is that NPV? For fractional years, assume that the annual costs are proportional to the fraction

TABLE 7.2

R&D Options

Project	Annual Cost ($ Millions)	Project Length (Years)	Potential Annual Profit ($ Millions)
A	1.0–5.0	2–4	10–16
B	2.2–4.3	3–4	12–15
C	6.5–8.7	4–6	16–24
D	4.4–6.0	5–7	15–45
E	5.0–6.5	5–8	21–29
F	1.1–2.0	1–2	4–6
G	5.8–8.0	4–8	11–24

of the year in which the R&D project is still ongoing. Annual profits are earned equally over 4 years after R&D is finished, regardless of whether it is a fractional or whole year. (*Hint:* it is probably easier to model the budget constraint as a requirement than as a decision constraint.)

b. Replace all of the certain variables with uncertain distributions. (Use PERT distributions, with the given values as the minimum and maximum values and the most likely value the midpoint of the range.) In which projects would you invest your fixed R&D budget of $75 million in order to maximize the expected NPV? What is that expected NPV? What is the probability that the eNPV will exceed $100 million?
c. Now suppose that the probability of project success is uncertain. Assume that the probabilities of successful R&D projects are .8, .75, .5, .9, .7, .85, and .75 for projects A through G, respectively. Repeat the questions in part (b).

7.6 FISHERIES MANAGEMENT

This exercise is based upon Exercise 5.6 in Chapter 5. Use the fish population biology model from that exercise, but modify the fishing fleet information as follows. Initially, each ship can catch 12 fish; this number will increase each year due to technological improvements, but also depending on the fish population. The additional catch each year is a decreasing function of population size, ranging from a catch of 12/ship for a population size of 4,000 and decreasing to a catch of 6/ship at a population size of 1,000. (Assume a linear relationship, based on these two points and rounded down to the next lowest integer.)

a. Given a fish price of $1,000/fish, find the number of fishing boats that maximizes the net present value of the catch (assuming a discount rate of 10%) over a 10-year time period. What is the probability of extinction in 10 years' time with the optimal number of boats? Repeat these questions for a 20-year time period.
b. Repeat part (a) with a discount rate of 1%.

TABLE 7.3

Component Failure Rates

Process	MTBF (Hours)	Cost for 10% Reduction in MTBF ($ Millions)
A	250	16
B	331	21
C	426	14
D	256	10
E	340	20
F	245	18
G	140	14
H	333	15
I	400	12

7.7 MULTIPLE COMPONENT FAILURES

Consider a manufacturing process with a number of activities that are arranged in a *serial* and a *parallel* configuration. Serial processes mean that if any earlier process fails, the entire process fails; parallel processes must all fail for the process to fail. This particular manufacturing process has the following chain: A, (B, C, D), E, F, (G, H), I. Letters contained within parentheses are parallel, while the others are serial. Table 7.3 provides data about the mean time between failures and the cost for improving the MTBF (used in part b) for each of these processes. Assume that each component's failure time can be modeled as an Exponential distribution, with the MTBF as the rate.

a. Build a spreadsheet to simulate the total time to system failure. What is the expected time to failure? Provide a 90% confidence interval.
b. Suppose that you can invest $100 million in improving the reliability of these processes. Table 7.3 shows the R&D cost (1-year duration with no uncertainty about success) for achieving a 10% reduction in the MTBF for each process. How should you spend the $100 million to achieve the greatest reduction in the expected time to system failure? (Partial projects are not feasible.) How should you invest the $100 million if you want to reduce the expected time to system failure as much as possible while guaranteeing that the 95th percentile of the time to system failure is at least 200 hours?

7.8 CONCERT PRICING

You are arranging a concert in a venue that seats 1,000 people. There are three tiers of seating, as described in Table 7.4.

a. Find the set of prices that maximizes total revenue from this event. Assume that if the seats in a particular tier are not sufficient to

TABLE 7.4

Concert Ticket Tiers

Tier	Seats Available	Demand Function
A	100	1,050 – 5*Price
B	400	20,400 – 250*Price
C	500	20,000 – 400*Price

satisfy demand, then the unsatisfied fans will not attend at all (i.e., they will not purchase a different tier's tickets; as an optional exercise, you can change this assumption so that unsatisfied demand will switch to the next lower priced tier if there is room available).

b. Suppose that you could reconfigure the venue to permit different proportions of seats in the first two tiers (but the third tier must have 500 seats). What is the revenue maximizing configuration and set of prices?

c. Now suppose that demand is uncertain. Table 7.4 provides the mean demand for each tier (using the linear demand as a function of price to provide the mean demand), but each demand is lognormally distributed, with standard deviations of 1,000, 80, and 100, respectively. (The standard deviations do not depend on the price.) What configuration and set of prices maximizes expected total revenue? What is the probability that revenue will not exceed $50,000?

d. Suppose that you wish to halve the probability of getting $50,000 (or less) of revenue. Is there a configuration of seats and prices that accomplishes this? If there is, what is the expected total revenue?

Appendix A: Monte Carlo Simulation Software

LEARNING OBJECTIVES

- Learn about different Monte Carlo simulation software packages for Excel.
- See the three basic steps of building a Monte Carlo simulation model in Excel spreadsheets.
- Develop a basic appreciation of some of the characteristics from common Monte Carlo simulation Excel add-ins: ModelRisk®, @RISK®, Crystal Ball®, and RiskSolver®.

A.1 Introduction

All of the screenshots, models, and solutions to the exercises within this book are provided and shown using the software package ModelRisk. In our work as consultants, trainers, and researchers, we use a variety of software packages. The choice of which package is best typically depends on a number of factors, including (1) the nature of the problems that you may try to simulate, (2) which package is used elsewhere in your organization or line of work, and (3) whether you already have been exposed to a particular package or not.

It is important to see the software package as just a tool, not as the solution. Often people interested in risk analysis equate the purchase of risk analysis software packages with "doing risk analysis." We hope that this book provides people interested in risk analysis with practical ideas and knowledge of techniques and methods independent of the specific software package you may use.

Having said that, there are many different software packages available (many more than the four discussed in this appendix) that vary in terms of their capabilities, features, and costs. Discussing all of those packages is well beyond the scope of this appendix. Instead, we will offer a brief overview of the four packages that are frequently used for Monte Carlo simulation.

A.2 A Brief Tour of Four Monte Carlo Packages

The four packages that will be discussed briefly are

- ModelRisk from Vose Software BVBA
- @RISK from Palisade Corporation
- Crystal Ball from Oracle
- RiskSolver from Frontline Systems

In general, all these packages provide the same basic Monte Carlo simulation functionality. In other words, they all allow the user to perform the three important steps for implementing an MC simulation:

- Step 1: build a Monte Carlo model and put probability distributions within cells in Excel.
- Step 2: run a Monte Carlo simulation for thousands of iterations.
- Step 3: review and present the results of the Monte Carlo simulation using a number of charts, etc.

Some key characteristics of the four packages include:

- *ModelRisk* from Vose Software BVBA:
  - Release of first version of ModelRisk in the mid-2000s
  - Comprehensive Monte Carlo simulation tool that contains a number of advanced and unique capabilities and techniques
  - Works with functions (similar to typical Excel functions) such as "VoseNormal (Mean, StDev)" or "VosePoisson (Rate)"
  - Relative smaller worldwide user base
  - Some availability of public training classes, plus custom classes can be requested
  - More information at www.vosesoftware.com
- *@RISK* from Palisade Corporation:
  - Release of first version in the second half of the 1980s
  - Comprehensive Monte Carlo simulation tool that is part of the DecisionTools Suite
  - Implemented in Excel via functions (similar to typical Excel functions) such as "RiskNormal (Mean, StDev)" or "RiskPoisson (Rate)"
  - Large worldwide user base in a great variety of industries and fields

- Lots of training classes, user conferences, case studies, etc. available
    - More information at www.palisade.com
- *Crystal Ball* from Oracle:
    - Release of first version of Crystal Ball in the second half of the 1980s
    - Comprehensive Monte Carlo simulation tool that can be integrated with several other Oracle tools (e.g., Essbase or Hyperion Strategic Planning)
    - Works with either functions or distributions (called "Crystal Ball Assumptions") that are not part of the native Excel environment (although the use of CB functions does not provide full functionality of the simulation outputs)
    - Large worldwide user base in a great variety of industries and fields
    - Fair amount of training classes (through Oracle University) available
    - More information at www.crystalball.com
- *RiskSolver* from Frontline Systems:
    - Release of first version of RiskSolver in the second half of the 2000s
    - Monte Carlo simulation tool with core strength in simulation optimization
    - Works with functions (similar to typical Excel functions) such as "PsiNormal (Mean, StDev)" or "PsiPoisson (Rate)"
    - Small worldwide user base, but used in a number of popular textbooks
    - No public training classes or user conferences
    - More information at http://www.solver.com/risksolver.htm

Finally, on the book website (http://www.epixanalytics.com/lehman-book.html), there is a link to *ModelAssist for Crystal Ball* and *ModelAssist for @RISK*, two free risk modeling training and reference tools that come with lots of example models.

Index

A

Absolute addressing, 9
Absorption state, 200
Adjusted R-Square, 151
ARCH models, *see* Autoregressive Conditional Heteroskedasticity models
ARMA models, *see* Autoregressive Moving Average models
Assumption(s)
  airline pricing problem, 212
  concert pricing, 250
  Crystal Ball, 253
  disaggregated, 82
  distribution function, 116
  economic growth, 92
  Efficient Market Hypothesis, 100
  electric vehicles, 18–19
  flu season, 206
  global warming, 2090
  historical patterns, 180
  Hollywood finance simulation, 50
  input, 13
  insurance model, 55, 60
  linear regression models, 150
  ModelRisk functions, 189
  Monte Carlo simulation, 235
  oil spills, 88
  parameter, retirement funds, 6, 7
  phone price, 2
  portfolio management, 246
  random sampling, 153
  rate of return, 25
  severity distributions, 81
  simplifying, 16
  snowfall at ski area, 89
  technology diffusion, 166–167
Auditing tools, 14, 16
Autoregressive Conditional Heteroskedasticity (ARCH) models, 188
Autoregressive Moving Average (ARMA) models, 188

B

Bayesian Model Averaging (BMA), 104
Bell-shaped curve, 113
Bernoulli distribution, 53, 111, 130
Best-case scenarios, modeling of, 26
Best practices, 15–17
Binomial distribution, 56, 64, 112
Black box, 228
Black Sawn events, 135
Blameless errors, 15
BMA, *see* Bayesian Model Averaging
Bootstrap
  forecast, 157
  procedure, 160, 161
  samples, 155
Box plot, 57

C

Capital Asset Pricing Model (CAPM), 91
CAPM, *see* Capital Asset Pricing Model
Cases, *see* Examples
CDOs, *see* Collateralized Debt Obligations
Central Limit Theorem, 65, 113, 218
Chart
  building of, 3
  cell phone pricing influence chart, 4
  influence, 3
  scatter, 148, 199
  Tornado, 42, 43, 47, 97, 133, 237
  trend, 180, 201, 205, 208
CLV, *see* Customer Lifetime Value
COGS, *see* Cost of Goods Sold
Collateralized Debt Obligations (CDOs), 134, 135
Conceptual maps and models, 1–22
  absolute addressing, 9
  auditing tools, 14, 16
  automated auditing tools, 15
  best practices, 15–18
  blameless errors, 15

business decision, 1
cell phone pricing
influence chart, 4
spreadsheet model, 2
chart, building of, 3
connector lines, 4
culpable errors, 15
Customer Lifetime Value, 20
Data Validation tools, 14
decision problem, typical, 1
documentation, 16
electric vehicles, 18–19
errors in spreadsheet modeling, 13–15
exercises, 18–22
formula testing, 16
good practices with spreadsheet model construction, 10–14
Handys, 1
health insurance choices, 18
Hollywood finance, 22
influence charts, 3
introductory case (mobile phone service), 1–2
investing in education, 19
lack of knowledge, 5
mobile phone service, 1
Monte Carlo simulation, 17
net present value, 19
network economics, 20–21
overconfidence, 17
parameter designation, 4
peak load pricing, 21
postretirement withdrawals, 9
profitability, measurement of, 22
project activities list, 22
project management, 21
qualitative errors, 15
quality control procedures, 17
quantitative errors, 15
relative addressing, 9
retirement fund visualization, 7
retirement planning example, 5–10
retirement spreadsheet, 8, 11
simplifying assumptions, 16
spreadsheet errors
identification, 14
research on, 13
spreadsheet model, value of, 2
visualization, 3–5
What If Analysis on toolbar, 12
Copula, 141
Correlation matrix, 139, 140
Cost of Goods Sold (COGS), 94
Crystal Ball Assumptions, 253
Culpable errors, 15
Cumulative distribution, 116
Customer Lifetime Value (CLV), 20

D

Decision making, *see* Optimization and decision making
Decision problem, typical, 1
Deduct distribution, 70
Deterministic fit, 104
Discounting, 91
Discrete distribution, 112
Discrete event models, 47
Distribution
Bernoulli, 53, 111, 130
Cumulative, 116
Discrete Uniform, 113
Exponential, 115
Gamma, 115
Lognormal, 115
nonparametric, 116
Ogive, 116
Pareto, 115
Relative, 116
Student, 115
Triangle, 116
Uniform, 115
Weibull, 115
Distributions, selection of, 91–128
airline delays, 124
alternative energy, 126–127
Apple, Inc., 92, 93, 97
average VaR, 110
Bayesian estimation, 117–122
Bayesian learning spreadsheet, 119
Bayesian Model Averaging, 104
bell-shaped curve, 113
Bernoulli distribution, 111
Binomial distribution, 112
broadband Internet usage, 125–126
capital adequacy, 99
Capital Asset Pricing Model, 91

Central Limit Theorem, 113
combined distribution, 97
commonly used continuous distributions, 113–116
commonly used discrete distributions, 111–113, 114
Cost of Goods Sold, 94
Cumulative distribution, 116
decision guide for selecting distributions, 116–117
deterministic fit, 104
discounting, 91
Discrete distribution, 112
Discrete Uniform distribution, 113
distribution fitting for VaR, parameter uncertainty, and model uncertainty, 99–111
Efficient Market Hypothesis, 100
exercises, 123–128
Exponential distribution, 115
first introductory case (valuation of public company), 91–93
First Order fit, 104
First Order uncertainty, 109
flat prior, 118, 121
free cash flows, 91
functional fitting, 116
Gamma distribution, 115
Gaussian distribution, 113
Geometric distribution, 112
Hypergeometric distribution, 112, 113
Information Criteria, 102
investment returns, 124
Laplace distribution, 102, 103
likelihood, 118
Log Likelihood, 105
Lognormal distribution, 115
log return, 100
modeling expert opinion in valuation model, 94–98
model uncertainty, ModelRisk representation of, 104
Monte Carlo simulation, 92
Monty Hall problem, 128
Negative Binomial distribution, 112
net cash flows, 91
non-Bayesian and Bayesian models, 122
nonparametric distributions, 116
number of defects, 113
Ogive distribution, 116
outliers, 109
overhead costs, 92
parameter uncertainty, 103, 104
Pareto distribution, 115
PERT distributions, 96
Poisson distribution, 112
Polya distribution, 112
posterior probability, 118
real estate crisis, 125
Relative distribution, 116
second introductory case (fitting distributions to data), 99
Second Order distribution fit, 104
Second Order uncertainty, 109
sensitivity analysis, 117
side effect probabilities, 121
smoking and birth weight, 123
SSRI inhibitors, 118
statistical theory, 118
Student distribution, 115
time series patterns, 100
Triangle distribution, 116
Uniform distribution, 115
valuation simulation results, 98
valuation tornado chart, 98
Value at Risk, 99, 108
warranties, 127
Weibull distribution, 115, 116
whiskers, 109

E

Efficient Frontier, 234
Efficient Market Hypothesis (EMH), 100
EMH, *see* Efficient Market Hypothesis
Empirical copulas, 143, 146–148
Error(s)
addressing, 9
blameless, 15
copy and paste, 15
costs associated with, 14
culpable, 15
entry, 131
Excel, 106, 148
identification, 14
internal rate of return, 50
message, 176, 201

prevention, 13
qualitative, 15
quantitative, 15
random sampling, 153
Standard, 152, 153, 160
Type I, 154
Type II, 154
Examples
drug development, 129–134
insurance problem, 51–52
mobile phone service, 1–2
retirement planning, 5–10, 23
Excel, 13
Add-ins, 151
array function, 80, 140, 192
built-in tools, 14
Data Analysis Regression tool, 162
Data Table command, 10
discrete event models, 47
error, 106, 148, 201
EXP function, 207
IRR function, 22
Monte Carlo, 28
scatter plot, 141
Scenario Manager, 25, 26
Show Formulas, 55
Solver, 221
STEYX function, 155
symbol button, 32
Vose functions, 33, 77
XIRR function, 50
Exercises
conceptual maps and models, 18–22
distributions, selection of, 123–128
Monte Carlo simulation, 48–50
objects, modeling with, 85–89
optimization and decision making, 244–250
relationships, modeling of, 165–169
time series models, 206–210
Expert-Opinion-Based models, 188
Exponential distribution, 115

F

Fast Fourier Transform (FFT) methods, 71, 73
FFT methods, *see* Fast Fourier Transform methods
First Order fit, 104
Flat prior, 118, 121
Free cash flows, 91
Functional fitting, 116

G

Gamma distribution, 115
GARCH model, 186, 187, 193
Gaussian distribution, 113
GBM models, *see* Geometric Brownian Motion models
Genetic algorithms, 220
Geometric Brownian Motion (GBM) models, 188
Geometric distribution, 112

H

Handys, 1
Hill climbing technique, 220
Hypergeometric distribution, 112, 113
Hypothesis
Efficient Market, 100
null, 152
rejection of, 153

I

IF statements, 132
Influence charts, 3, 14
Information Criteria, 102

J

Jump diffusion, 188

L

Littlewood's Rule, 212, 216, 218, 228
Log Likelihood, 105
Lognormal distribution, 65, 115
Log return, 100
Long strangle, 206
Long-Term Capital Management (LTCM), 135
Lotus, 13
LTCM, *see* Long-Term Capital Management

M

Maps, *see* Conceptual maps and models
Memoryless system, 200
Mobile phone service, 1
Model, *see also* Conceptual maps and models; Time series models
  AggregateFFT, 74, 75
  airline pricing, 213
  ARCH, 188
  Capital Asset Pricing, 91
  CDO default, 136
  construction, good practices, 10
  discrete event, 47
  drug development, 131
  Expert-Opinion-Based, 188
  GARCH, 186, 187, 193
  insurance, 55, 60
  linear regression, 150
  Markov, 220
  multiple damages, 79
  multiple linear regression, 160
  multivariate time series, 196, 198
  Object-based, 65
  oil price, 195
  retirement planning under uncertainty, 41
  uncertainty, ModelRisk representation of, 104
  variability/uncertainty, 83
  visualization, 13
ModelRisk, 29, 30, 252
  Aggregate functions, 61, 71, 73
  Combine option, 96
  copula structures, 142
  Data Viewer, 190
  discrete distributions, 111
  distribution fitting, 142
  Distribution Object in, 76
  Empirical copula, 146
  Expert tool, 94, 95
  Information Criteria, 102
  insurance modeling, 69
  Markov Chain modeling, 201
  model uncertainty, representation of, 104
  Monte Carlo simulation using, 30–38
  Objects, 59
  OptQuest, 221, 222
  Output range, 180
  SimTable functionality in, 42
  Simulation Settings window, 217
  Stats display, 55
  stochastic dominance, 238
  time series model, 175, 176, 189
  trend charts in, 186, 204
  Univariate Time Series, 175
  U parameter in, 138
  VoseBinomial functions, 60
  VoseSimProbability function, 77
Monte Carlo simulation, 17, 23–50, 92
  Ascending Cumulative Plot view, 37
  best-case scenarios, modeling of, 26
  cell phone pricing model
    probability distributions for, 30
    revised, 33
  common mistake in, 173
  contract length distribution, 32
  cumulative probability functions and random number generation, 29
  customer lifetime value simulation, 49
  discount rate, 33
  discrete and continuous probability distributions, 25
  discrete event models, 47
  educational investment simulation, 48–49
  electric vehicle cost simulation, 48
  employee contribution, 42
  exercises, 48–50
  health insurance choice simulation, 48
  historical stock market data, 39
  Hollywood finance simulation, 50
  internal rate of return, 50
  introductory case (retirement planning), 23
  Mark Output/Input dialog box, 36
  ModelRisk, 30–38
  Monte Carlo simulation, 27–30
  National Credit Union Administration, 46
  negligible probability, 27
  network economics simulation, 49
  optimistic scenarios, modeling of, 26
  overlay results, 38
  peak load pricing simulation, 49

PERT distribution, 29, 33, 39
pessimistic scenarios, modeling of, 26
probability distribution, 24, 25
project management simulation, 49–50
random-number generator, 47
retirement planning, 39–47
risk and uncertainty, 24–25
Scenario Manager, 25, 26
Select Distribution, 30, 31, 32
SimTable, 44, 45, 46
Simulation Settings dialog box, 36
software, 28
Tornado chart, 42, 43
View Function button, 35
worst-case scenarios, modeling of, 26
Monte Carlo simulation software, 251–253
Crystal Ball, 253
implementing steps, 252
ModelRisk, 252
Monte Carlo packages, 252–253
package characteristics, 252
@RISK, 252
risk analysis software, 251
RiskSolver, 253
Moral hazard, 52
Multiple R, 151
Multivariate copula, 144, 145
Multivariate time series model, 196, 198

N

National Credit Union Administration (NCUA), 46
NCUA, *see* National Credit Union Administration
Negative Binomial distribution, 112
Net cash flows, 91
Net present value (NPV), 19
drug development, 131–133, 165
Hollywood finance simulation, 50
investing in education, 19, 48
project selection, 247
scenarios, 237
Nonparametric distribution, 116
Normal distribution, 65
NPV, *see* Net present value
Null hypothesis, 152

O

Objects, modeling with, 51–89
AggregateFFT model, 74, 75
automobile insurance, 77, 86
Bernoulli distribution, 53, 54
Binomial distribution, 56, 64
box plot, 57
Central Limit Theorem, 65
Claim Amount, 57
Claim Frequency distribution, 64
customer loyalty programs, 86–87
Deduct distribution, 70
driver formulas, 61
Excel array function, 80
exercises, 85–89
extended warranties, 52
Fast Fourier Transform methods, 71, 73
frequency of losses, 52
frequency and severity, 52–59
hospital bed utilization, 85
insurance model
basic, 55
one-row, 58, 59, 63
troubleshooting, 60
insurance policy types, 86
insurance premium results, 56
insurance visualization, 53
Internet spam marketing, 87
introductory case (insurance problem), 51–52
lognormal distributions, 53, 54
modeling deductibles, 69–76
modeling frequency/severity without using objects, 65–68
Monte Carlo Aggregate simulation, 71
moral hazard, 52
multiple damages model, 79
multiple severity/frequency distributions, 77–82
no correlation case, 80, 81
non-claims-related expenses, 77
Normal distribution, 65
objects, 59–60
oil spills, 88
Paste to Excel button, 60

Poisson distribution, 64
probability of success, 53
probability theory, 56, 65, 68
required premium, 69
severity of losses, 52
SimTable, 72
simulation output cells, overlaying of, 56, 59
snowfall at ski area, 89
standard policy, 77
supermarket checkout, 88–89
true parameter value, uncertainty about, 82
uncertainty and variability, 82–85
unrealistic situation, simulation of, 57
using objects in insurance model, 60–65
using objects without simulation, 76–77
utility company repair vehicles, 85
variability/uncertainty model, 83
whiskers, 57

Ogive distribution, 116
Optimistic scenarios, modeling of, 26
Optimization and decision making, 211–250
adding requirements, 231–234
airline pricing model, 213
black box, 228
Central Limit Theory, 218
complex objective function, 221
concert pricing, 249–250
cumulative distribution functions, 240
decision variables, 231
deterministic simulation results, 225
deterministic solution analyses, 226
distribution of optimal solution, 234
Efficient Frontier, 234
empty seat results, 217
exercises, 244–250
expected incremental revenue, 212
expected outcome, 237
fisheries management, 248
four weeks of formulas, 215
fuel costs versus labor costs, 245–246
future scenarios, 237
genetic algorithms, 220
graphical displays, 237
hill climbing technique, 220
histogram, 237
Initial Solution Analysis with requirement, 236
introductory case (airline seat pricing), 211–212
Littlewood's Rule, 212, 216, 218, 228
mean demands, 230
modeling prices, 230
multiple component failures, 249
no-control strategy, 216
optimal inventory management, 246–247
optimization with multiple decision variables, 229–231
optimization with simulation, 226–228
optimization software, 219
optimization solution to airline pricing problem, 219–225
OptQuest
decision window, 223
settings window, 224
target window, 222
portfolio management, 246
presentation of model results, 237
presenting results for decision making, 235–237
probability density functions, 240
project selection, 247–248
Q class passengers, 212, 226
quality-adjusted life years, 244–245
requirement addition, 235
revenue results for three strategies, 216
risk reduction, 233
risk–return trade-offs, 234
simulation model of airline pricing problem, 212–216
Simulation Requirements, 232
simulation table, 216–218
results, 218
setup, 217
Solver, 221
stochastic dominance, 238–244
stochastic optimization

results, 228
settings, 227
Stochastic Solution Analysis, 229
summary, 244
truncated distribution, 214
variable requirement, 234
VoseDominance array functions, 242
Overconfidence, 17

P

Parameter uncertainty, modeling of, 104
Parametric Bootstrap, 155
Pareto distribution, 115
PERT distribution, 29, 33
cell phone pricing, 29
customer lifetime value simulation, 49
demand sensitivity parameter, 29
drug development, 130
Envelope method, 163
Hollywood finance simulation, 50
hospital bed utilization, 85
Internet spam marketing, 87
ModelRisk functions, 33, 96
Monte Carlo simulation, 39
supply shock, 194
time series version, 189
Triangle distribution, 116
Pessimistic scenarios, modeling of, 26
Pivot table, 177
Poisson distribution, 64, 112
Polya distribution, 112
Posterior probability, 118
Power Laws, 169
Probability distribution, 24
Probability of success, 53
P-value, 153, 154

Q

Qualitative errors, 15
Quality control procedures, 17
Quantitative errors, 15

R

Random-number generator, 47
Random walk models, 188
Regression analysis, 149
Relationships, modeling of, 129–169
Adjusted R-Square, 151
advertising–sales relationship, 155
airfares, 165–166
analyst judgment, 154
arrow labels, 150
Bernoulli distributions, 130
Black Sawn events, 135
Bootstrap
forecast, 157
procedure, 160, 161
samples, 155
box plots, 157, 158
Coefficient of true model, 153
Collateralized Debt Obligations, 134, 135
copulas, 141
correlation matrix, 139, 140
covariance, 137
default rates, 135
drug development, 165
Empirical copulas, 143, 146–148
Envelope method, 161–163
errors, types of, 141
Excel error, 148
exercises, 165–169
expected NPV, 131, 133
first example (drug development), 129–134
fish population, 168
fourth example (advertising effectiveness), 148–149
home price changes, 142
hurdle rate, 133
hypothesis test, alternative to, 153
hypothetical advertising and sales data, 149
IF statements, 132
income and housing, 166
judgment, 154
logical relationships, 129
Long Tail, 169
Long-Term Capital Management, 135
multiple correlations, 139–141
Multiple R, 151
multiple regression models, 159–161

Multivariate copula, 144, 145
null hypothesis, 152
parametric Bootstrap, 155
portfolio correlation, 166
Power Laws, 169
probability of default, 137
P-value, 153, 154
random sampling error, 153
regression
analysis, 149
assumption, 164
modeling, 150–154
R-Square, 151
scatter chart, 148
second example (Collateralized Debt Obligations), 134–139
Significance F, 152
simulation within regression models, 154–159
Standard Error, 152, 153
statistical theory, 153
summary, 164–165
technology diffusion, 166–167
third example (copulas), 141–145
Tornado chart, 134
Trendline equations, 163
U parameter (ModelRisk), 138
warranties, 168–169
Zipf's Law, 169
Relative addressing, 9
Relative distribution, 116
Revenue passenger mile (RPM) data, 171
Risk analysis software, 251
RPM data, *see* Revenue passenger mile data
R-Square, 151

S

Scatter chart, 148, 199
Scenario Manager, 25
Second Order distribution fit, 104
Select Distribution, 30
Sensitivity analysis, 117
Significance F, 152
SimTable, 44, 45, 46, 72
Simulation Requirements, 232
Software, *see also* Excel; ModelRisk; Monte Carlo simulation software
discrete event simulation, 47
Monte Carlo simulation, 28, 30, 96
numbers generation, 28
optimization, 219
purpose, 28
risk analysis, 251
statistical, 159
Solver, 221
Spreadsheet model, *see* Model
Standard Error, 152, 153, 160
Statistical software, 159
Stochastic dominance, 238–244
Stochastic Markov Process, problems modeled as, 200
Strike price, 207
Student distribution, 115

T

Time series models, 171–210
absorption state, 200
airline load factors, 206
analysis of air traffic data, 175–183
ARCH models, 188
backcasts, 183
confidence interval, 172
deseasonalized and detrended data, 178
Detroit versus New York simulation results, 199
e-commerce and retail sales, 208
error message, 176, 201
exercises, 206–210
Expert-Opinion-Based models, 188
fit time series dialog box, 179
flu season, 206
fourth example (home prices and multivariate time series), 196–199
GARCH model, 186, 187, 193
Geometric Brownian Motion models, 188
global warming, 209
hybrid oil price model, 195
hypothetical data set, 173, 174

introductory case (September 11 and air travel), 171–172
jump diffusion, 188
long strangle, 206
Markov chains, 200–204
Markov model, 202, 203
mean, 172
memoryless system, 200
multivariate city house price data, 197
multivariate time series model, 196, 198
need for time series analysis, 172–174
nested logical statements, 201
new car sales, 209
Object with last known value, 180
oil price trend chart, 193
Olympic swimsuit controversy, 209–210
OPEC behavior, 190
patents, 208
pivot table, 177
random walk models, 188
range names, 201
removing seasonality and trend, 178
revenue passenger mile data, 171
scatter chart, 199
seasonality, 171, 176
second example (stock prices), 183–187
simulated air traffic, 182
Stochastic Markov Process, problems modeled as, 200
strangle options, 206–208
strike price, 207
summary, 205
third example (oil prices), 190–195
time series data, 173
transition matrix, 201
trend charts, 180, 205, 208
types, 188–189
U.S. air passenger travel, 172

Tornado chart, 42, 43, 47, 97, 133, 237
Trend charts, 180, 201, 205, 208
Trendline equations, 163
Trial and error, 219
Triangle distribution, 116

U

Uncertainty
effects of on average insurance policy cost, 84
First Order, 109
gambling casino, 52
model, ModelRisk representation of, 104
modeling of, 236
parameter, modeling of, 104
risk and, 24–25
Second Order, 109
true parameter value, 82
visualization for retirement planning under, 40

Uniform distribution, 115
U parameter (ModelRisk), 138

V

Value at Risk (VaR), 99
VaR, *see* Value at Risk
VisiCalc, 13
Visualization, 3–5
insurance, 53
model building using, 13
profitability, 22
retirement fund, 7
retirement planning under uncertainty, 40
time structure, 6

W

Weibull distribution, 115, 116
What if analysis
capability of spreadsheet model to perform, 2, 10
expected NPV, 133
Scenario Manager, 25
toolbar notation, 12

Whiskers, 57, 109
Worst-case scenarios, modeling of, 26

Z

Zipf's Law, 169